THE SPACE ORACLE

THE SPACE ORACLE: *A Guide To Your Stars*
By Ken Hollings
First published by Strange Attractor Press in 2018,
in an unlimited paperback edition.
Text © Ken Hollings 2018

Design by Tihana Šare.
Typeset in Crimson Text.

ISBN: 978-1907222-535

All images taken from MS.2270, Wellcome Collection, London

Distributed by The MIT Press, Cambridge, Massachusetts.
And London, England.

Printed and bound in Estonia by Tallinna Raamatutrükikoda.

STRANGE ATTRACTOR PRESS
BM SAP, London, WC1N 3XX, UK
www.strangeattractor.co.uk

THE SPACE ORACLE

A Guide to Your Stars

Ken Hollings

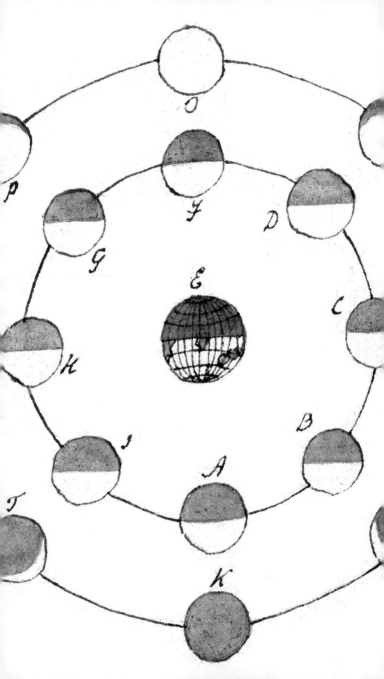

CONTENTS

CELESTIAL NAVIGATION

How To Use This Book

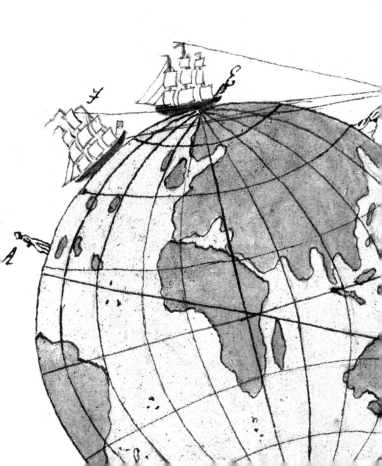

I Sekuin, Perfected These Arts
Along The Streets Of Minraud.
Under Sign Of The Centipede.

William Burroughs,
The Soft Machine

The horoscope for John Milton 'borne the 9th of December 1608, *die Veneris,* half an hour after 6 in the morning, in Bread Street in London', was cast around 1650 by 'John Gadbury: the Astrologer'. Despite our best intentions to know better, the astrological structures set down by the astronomer Claudius Ptolemy almost two thousand years ago have remained embedded in our culture. Even if we don't believe in what was once hailed as the Art of Kings, we are each secretly familiar with our individual star sign. Astrology is historically intertwined with the development not just of astronomy but also mathematics, chemistry, medicine, music and literature. As well as casting horoscopes, John Gadbury also compiled almanacs for farmers and merchants. Bodies, lands and principalities were revealed through the astrologer's skills as things that existed in time as well as space – everything on Earth had a point of temporal origin, which was also a form of fate. When construction began on the Royal Observatory at Greenwich in London, the Astronomer Royal John Flamsteed cast its horoscope based on the precise date and time when the foundation stone was laid: 3.14 pm on 10 August 1675. The chart drawn up by Flamsteed included a quote from the Latin poet Horace: '*Risum teneatis amici*' – 'could you stop yourselves from laughing, friends?'

The Space Oracle is no more a defence of astrology than it is an attack on astronomy, but rather a critical examination of how such complementary systems organise experience. Both offer us valuable examples of how knowledge is structured in a manner that renders it indistinguishable from certainty.

Meanwhile the cosmos remains a work in progress – and this book seeks to examine how fragmented this process actually is. Space is an assemblage of overlapping perspectives, fragments of old movie stills, comic books, instrumentation, charts, maps and toys. Cosmology is, as a consequence, not so much about creating an integrated whole but of separating out our perceptions into an orderly pattern. Every cosmos we have devised so far begins and ends as a fiction and, as such, can never be rendered entirely coherent. The twelve houses of the Zodiac, literally a circle of animals looking down at us from the night sky, still serve this purpose.

I have therefore retained the order and sequence of the old Ptolemaic astrological calendar in the structuring of this book but have changed the designations of the individual houses – not out of any wilful desire to be obscure but to remind myself that names are only customs, which means they represent a habit of mind not a universal principle. The Babylonians, the Ancient Egyptians and the Maya would not have recognised 'Mars' or 'Sirius' or 'Venus', but they still charted the movements of these points of light across their skies with great care and ascribed certain powers to them.

Similarly, the names attributed to our astrological houses are equally arbitrary and therefore also subject to change. Just as today's star charts and maps of planets are periodically updated to reflect new discoveries or changes of status, so the Zodiac might be radically revised to reflect a far longer, deeper and more complex history. Nicholas of Lynn, the fourteenth-century astronomer, published a *Kalendarium* of detailed tables that linked human health to the motions of the stars. 'If anyone however has an almanac and wants to get closer to the truth,' Nicholas wrote, 'he ought to calculate all the planets for the hour at which the illness began, and to place their calculated longitudes in the houses which he has calculated.' This diagnostic approach was intended to reflect a series of correspondences established between the human body and the signs of the Zodiac. Starting with the head and ending with the feet, an anatomical calendar extended from organ to organ: a concordance that I have attempted to retain in the final ordering of *The Space Oracle*. This means that you can look up your new star sign according to your date of birth and discover what parts of your body it rules and other related areas of influence. The twelve astrological houses that make up this book have been broken down further into five numbered sections. This is as a reminder that it was the Babylonians, with their clay tablets and reed pens, who first taught us to count in multiples of sixty, a process that involves so many integers – it's thanks to these early reckoners that there are sixty seconds in a minute, sixty minutes in an hour and six times sixty degrees in a circle. The distribution of the five sections across each of the twelve houses also means that the reader is free to read them in any numerically determined order they choose.

When it came to *The Space Oracle*'s content, I wanted to create something that combined the factual approach of a school textbook or an astronomical guide with the eclectic mysteries of early alchemic texts, which were quite often hotchpotches of astrology, chemistry, records of eclipses, experiments, spells and recipes. In many ways these early texts resembled the published archives I was examining while researching this book: loose collections of broadsheets, postcards, newspaper clippings or outdated atlases of the Moon. I also became fascinated by 'space age' books published for children over the past sixty years, filled with planets and rockets and cartoon drawings of experiments with magnets and light bulbs. Similarly the more recent textbooks I examined offered generalised perspectives on the arcana of celestial mechanics: facts and formulae organised to establish an orderly portrait of a seemingly unknowable chaos.

As with the best art and the boldest propositions, certainty suddenly confounds itself. During my researches

I kept coming back to specific questions. What constitutes a textbook? What gives it authority? How do we shape and communicate knowledge? Is there an art to it? What, in the testing of that knowledge, do we gain and what do we lose? What did we know in the past that we seem to have forgotten now?

The organisation of this book seeks to present a kaleidoscope of perspectives: this simple optical instrument, whose randomly generated patterns connect it directly to the visual design process, also reveals the effect that pieces of information, and the order in which they appear, have upon each other. Not surprisingly, Joseph Cornell's final series of boxes, popularly known as the 'Celestial Navigations', haunted my thinking all through the writing of this book. First exhibited in New York around 1957, these assemblages of 'space objects' brought together old astronomical charts, blue marbles, white sand, cordial glasses, shells and ball bearings into formations of mute arcana serenely contained between wooden walls. For me, Joseph Cornell was always the most literary of artists, not because of his references or allusions but purely because of his methods – to look at one of his neatly balanced and composed works is to understand how Cornell made art like a true writer.

I have paid formal homage to him in the naming of this book, which owes its origin to *The Ladies' Oracle*, a volume of parlour divination supposedly authored by the great alchemist Cornelius Agrippa. This small book made a great show of proffering essential assistance to anyone anxious to learn more about a romantic entanglement. The idea that a reader could interact with a printed and bound volume in order to find out about their love life appealed greatly to me. Astrology books, like textbooks and dictionaries, exist in a similar manner to be consulted. Knowledge, it seems to me, is a function completed by the agency of the reader – or rather is activated by it.

Light pollution spreading over our cities has transformed astronomy into an online experience. Our computers have become our observatories, and space increasingly inhabits the network as a timeline of stars and planets. Such an intense data storm of space-related texts, images and films blows through our social media that any response is determined by pattern recognition. *The Space Oracle* is an attempt to discover the shape and dynamics of that pattern and to trace them back to the same night sky that our distant ancestors first gazed upon thousands of years before we ever did.

Ken Hollings
London 2018

SCARAB

21 March – 20 April

SCARAB

21 March – 20 April

Also known as:
Scarabaeus satyrus or
The Dung Beetle

Body Part:
Head

Areas of Influence:
Perception, Alignment and
Path Finding

I

Followed a flight of stairs down to the eye surgery ward, where air conditioning units purred outside shadowy, enclosed windows. A bespectacled German nurse appeared with a folded hospital garment in her hands. 'Ah, you are already in your gown, isn't it?' she observed. 'But you must walk to the operating area. Then come with me. You don't have your wristband? Okay we will give you another one. Is your name still Kenneth Hollings?'

We walked along a long bright corridor through a set of glass doors to be met by the anaesthetist. He led me down another long corridor with high windows on one side, late-afternoon sunlight flooding in from the courtyard outside.

'Always the end one,' he said, nodding towards the operating theatre door.

'Oh I don't mind. The light looks so beautiful.'

'It's been a glorious day,' he replied.

He sat me down on the operating table and asked me to lie back. They shoved a pillow under my knees, fed a tube into the back of my hand, stuck on some ECG patches, covered me in a blanket then wheeled me through to theatre. A huge optical device like the interior of a small planetarium appeared above my head. I laughed. The sedative had already started to kick in. They wrapped blue plastic sheeting around my upper body, going all of the way up until it completely covered the right side of my face. Now completely blind, I felt my forehead and cheek being washed down with some kind of solution, but they may as well have been cleaning off the head of a marble statue for all I knew.

The local anaesthetic killed all sensation. I recall the operation only as an uninterrupted succession of sounds – heavy buzzes, high-pitched crackles, sizzles and trickles resonated inside my skull, creating a series of rippling acoustic spaces around me: the most amazing music I have ever heard. There were flickering lights and snatches of gentle conversation. Talk of lasers and sutures. 'No,' the surgeon said at one point. 'Leave those instruments in his eye.'

I felt a sudden pain and flinched.

'Are you all right?'

'That felt sharp.'

'Give him more morphine. Is that any better?'

'Still hurts a little.'

'Can I proceed?'

'Yeah, do it.'

'Not if it's still hurting – there's no point. Can you still feel anything?'

'Not now.'

Finally the blue plastic wrap was removed, my eye was taped up, and I was wheeled back out and told to sit up slowly. A member of the theatre team helped me into a wheelchair then pushed me through to post-op.

'I can go faster if you like,' she said.

'No, that's OK. I want to enjoy the evening light.'

II

The African dung beetle travels in straight lines at night, navigating by the light of the Milky Way. Astronauts operating outside the protection of Earth's magnetic field frequently see mysterious flashes in the darkness of space: members of the various Apollo missions, including Neil Armstrong and Buzz Aldrin, have reported seeing them. They have been described as resembling spots, streaks or clouds and frequently appear to be either white or colourless.

The human senses serve as a placeholder for experience. The range of human responses – what we see, hear and touch – is only one small part of a spectrum of activity that shades into the unseen, the unheard and the unfelt. Something like 70% of the cerebral cortex is used for seeing: the human brain automatically processes the world visually. The eye makes maps according to where the light is coming from. Atop its rolling ball of dung, the African dung beetle follows lines of polarised starlight, its shiny black shell reflecting bright symmetrical patterns.

The flashes seen by astronauts are cosmic rays streaking through the fluid inside their eyeballs: high-energy subatomic particles leaving bright, momentary trails behind them. 'There is a radiation hot spot in orbit,' according to NASA astronaut Don Pettit, 'a place where the flux of cosmic rays is 10 to 100 times greater than the rest of the orbital path. Situated southeast of Argentina, this region (called the South Atlantic Anomaly) extends about halfway across the Atlantic Ocean. As we pass through this region, eye flashes will increase from one or two every 10 minutes to several per minute.' According to a NASA survey, at least thirty-six former astronauts have developed cataracts within ten years of taking part in a high-radiation space mission. Digital cameras on board the International Space Station can be damaged so much by cosmic rays over the course of a year that they are no longer capable of taking a clear picture. A white snow of displaced pixels slowly accumulates on each image. The effect is permanent.

III

The dung beetle rolls its ball of excrement along a straight line until it finds a patch of soft sand in which to dig a hole. Rolling the dung into the hole, it then lays its eggs. Eventually the larvae will hatch out, devouring the ball as they grow. There is simply no way to make this sound more interesting than it is – except to note that the Ancient Egyptians venerated the sacred scarab as an agent of spontaneous generation able to produce living creatures directly from dead matter.

Dust and decay, dry particles existing on the edge of nothing, bring forth new life. The Sun shrinks before our eyes, losing 4.26 kilograms of its mass as energy with every passing second. Suddenly the desert floor teems with a fully formed brood of scarab beetles, born and raised out of excrement. The scarab-headed god Khepri of the Ancient Egyptians is therefore an aspect of the sun god Ra and the embodiment of both the dawn and this world's creation. The scarab hieroglyphic forms the first part of Khepri's name, which means 'He Who Is Coming Into Being'. Located firmly in the present tense, his emergence is depicted as something eternal and that can therefore never be completed. A black scarab squats upon Khepri's face, hiding his human features behind the impassive symmetry of its shell, legs and antennae.

The eyeball is a secret chamber in the head where light goes to die. Sight is planted and grows in the brain.

The dung ball is a blind eye, forcibly expelled from the body and buried in a socket of dry earth to await – what? This process is so fundamental that it is hardly noticed. With the start of each new day Khepri rolls the Sun like a ball of dung in a straight line across the horizon. The African dung beetle navigating by a galaxy's polarised radiation can be interpreted as just another aspect of the same basic activity. It simultaneously connects excrement, blindness and distance with a power so great that it seems alternately divine, horrifying and cruel.

IV

'*No,*' the surgeon said at one point. '*Leave those instruments in his eye.*' The shock of these words only struck some weeks later, their implications spreading like a wave of panic through my body.

Hanging in deep space, NASA's Kepler telescope captures the flash of an exploding star. What astronomers call the 'shock breakout' occurs when plasma jets cut through the star's outer layers after its core collapses. Before it suddenly broke apart, the red supergiant KSN 2011d was approximately five hundred times the size of the Sun and over one billion light years away from Earth. Sometimes there is *only* poetry in numbers: they make no sense any other way. All the heavy elements in the universe – including all of the iron, gold, silver and platinum found both in the earth and in our bodies – come from supernova activity. Life starts with stars exploding.

The stars are still there when we close our eyes. Beneath a darkening sky someone has written 'KEEP GOING STARS' in big white chalk letters on the path outside, the branches of tall trees hanging like shadows overhead.

V

The classification of colour reveals the madness of the rational mind: the visual spectrum only gains predominance once the Sun has been safely shifted to the centre of our Solar System. 'One might add,' observes Georges Bataille in the early part of the twentieth century, 'that the sun has also been mythologically expressed by a man slashing his own throat, as well as by an anthropomorphic being *deprived of a head*.' Sir Isaac Newton postponed publication of his *Opticks* until Sir Robert Hooke, the fiercest critic of his theories on the composition of light, had been dead for a year. The act of cutting and dividing had become an established form of inquiry by then: which is to say that the pursuit of knowledge is the systematic insertion of distance into experience. The eighteenth-century eye almost tore itself apart in its attempts to understand colour.

Any and all ordering of the subject is purely an illusion. Names and groupings drift like phantoms deep inside the eye. The meaning of the word 'spectrum' is historically associated with ghosts, apparitions and images: caught between geometry, optics and the act of seeing, its common usage refers simply to the resolved appearance of something unresolved. '*My Design in this Book is not to explain the Properties of Light by Hypotheses, but to propose and prove them by Reason and Experiments*,' Newton writes at the beginning of his 'Treatise of the Reflections, Refractions, Inflections and Colours of Light', first published in 1704. He had set out an arrangement of prisms, little more than toys bought at travelling fairs, in a darkened chamber to prove that light was not a wave but actually 'corpuscular' in nature. Newton also demonstrated that it can be separated into its constituent colours, which he determined to be seven in number, thus aligning them with the seven notes of the Pythagorean musical scale, the seven known objects in the Solar System and, by extension, the Seven Days of Creation. Spectres and shades drift through the room, recombining as white light on the far wall. Newton sets his seven colours in a divided black ink circle drawn on plain paper: they are Red, Orange, Yellow, Green, Blue, Indigo and Violet.

The eighteenth-century eye, following Newton, organised light into a wheel of twelve precisely connected colours. In one harmonious round, the primary colours Red, Yellow and Blue combine to form the three secondary colours Orange, Green and Violet; and between them emerge the six tertiary colours whose designations may vary but whose composition remains evenly regulated: Green-Yellow, Yellow-Orange, Orange-Red, Red-Violet, Violet-Blue and Blue-Green. Lenses and prisms transform light into a well-regulated machine: the eye, in turn, is trained to become an optical instrument. The Colour Wheel, elegantly engraved and hand retouched, is set within a trim landscape featuring a palace in the background, shrubs and an urn in the foreground, plus a female courtier admiring herself in a mirror held by Cupid. Like the eighteenth-century pastoral, the Colour Wheel is a correction of Nature – a form of perspective established across light instead of space.

Light is optical radiation: the spectrum of colour runs from ultraviolet at the short end of the waveband to infrared at the long end. Radio is very red light – redder than Infrared. Infrared radiation is detected by human skin as the sensation of heat; ultraviolet radiation manifests as a darkening of the skin. Spectrographic images render things visible for an eye that has become increasingly removed from the subject. The individual spectrum of light emitted by each star or galaxy is vital to identifying its unique physical properties within that range: beyond that, however, the colours ascribed to the image remain purely arbitrary. Outside the sunlit world of human sight and human understanding, the universe looks like nothing at all. The data streams from the Hubble Space Telescope, the Cassini probe orbiting Saturn or the New Horizons probe travelling beyond Pluto to the edge of the Solar System, all come back to Earth in raw black and white. Colours become wavelengths to be filtered out, captured in black and white and then composited together. Where does darkness lie on the spectrum – what wavelength does it occupy? Darkness is the density and texture of space. Transmitted as a sequence of zeroes and ones, the radiation emitted by distant galaxies is rendered visible as part of a universal Colour Wheel located at the back of the eye.

IXTAB

21 April – 21 May

IXTAB

21 April – 21 May

Also known as:
The Rope Woman or
The Goddess of Suicides

Body Part:
Neck and Throat

Areas of Influence:
Sacrifice, Grace and Fertility

I

'We introduce ourselves to planets and flowers,' observed Emily Dickinson – an exchange of greetings that indicates the degree to which all knowledge starts with the establishing of formalities. There is, for example, the story of a people who risked causing great offence to the stars during a lunar eclipse by banging pots and pans together and yelling into the dark night in order to keep the fading Moon awake. Astronomy consequently introduces an etiquette that requires several degrees of precision; and a proper study of the cosmos demands the highest order possible. In grouping planets and flowers together, scattered as they are in random profusion across the human field of vision, Dickinson establishes a vertical link between the Earth and the sky. Many mythologies place a great tree at the centre of the world: an *axis mundi* connecting the Earth with the Underworld and with Heaven. According to Norse Cosmology, *Yggdrasil*, the World Ash Tree, joins together the nine realms of the created universe. The ancient Maya had *Yaxche*, a mythical Ceiba tree with buttressed roots reaching down into Hell, a smooth trunk rising from the Earth to a huge spreading canopy in the sky, its branches indicating the four cardinal points of the horizon. Bats dwelt among *Yaxche*'s lower reaches while eagles nested in its branches.

Earthly creation is frequently portrayed as a form of descent. Nikolai Rynin, an early Russian proponent of space exploration, offered unsubstantiated accounts of the following legends: that the Chinese believed their distant ancestors fell to this world from the Moon; that Manco Guella, founder of the first Peruvian dynasty, came down from Heaven with his wife; and that the gods of the ancient Maya lowered themselves to Earth on a spider's web. To descend from the sky is to fall into division – into everything that separates the earthly from the celestial, the human from the divine, order from chaos, dream from conscious thought, male from female, subject from object. It is also the descent into distance and formality – into a world disrupted by space. Divided against themselves and from each other, the first humans are subject to the laws of language. In 1970

NASA published an English translation of Nikolai Rynin's nine-volume *Interplanetary Flight and Communication*, the first encyclopaedia on the history and theory of human spaceflight, in association with the Israel Program for Scientific Translations in Jerusalem.

All knowledge is built upon error – assumptions are made then modified or discarded. What remains unchanged and unquestioned is the formality of the inquiry itself. Distance from the subject is still the guarantee of certainty. In social terms, this manifests itself as patriarchy – the arrival of kings, priests and chiefs. Language and law thrive on division.

II

The priests arrive:

By day our Heaven is sky and light – by night it becomes darkness and stars. We understand more than those who come after us will ever know. Standing in the dark at night with the world beneath our feet, we turn the stars into gods. According to our eyes and senses, the cosmos starts with this. We need darkness in order to see the stars.

Stars connect us to the Earth. Did the Sun come before our kings? Did the Moon follow after our queens? First we had to stand up straight and raise our heads in order to bow before the stars.

Priests and farmers are the earliest astronomers. The Sun and Moon regularly preside over fertility and famine. Tides and seasons follow each other, leaving flooded plains and burning fields behind them. Observation and measurement form the basis for all prediction, which gives us our instinctive sense of eternal order – of the underlying and unfailing laws that will never change.

III

The ancient Maya originated in the Yucatan Peninsula over four thousand years ago. One of the oldest Meso-American civilisations, they developed an extensive hieroglyphic script and the only Pre-Columbian form of written language. 'The Mayan writings have not been fully deciphered,' William Burroughs noted in *The Soft Machine*, 'but we know that most of the hieroglyphs refer to dates in the calendar, and that these numerals have been translated – It is probable that the other undeciphered symbols refer to the ceremonial calendar.' The ancient Maya also developed sophisticated systems of astronomical observation and prediction, practiced an early form of slash and burn agriculture and performed ritual human sacrifice – traditionally by decapitation, removal of the heart from a living victim or by piercing the body with arrows.

The Dresden Codex is the oldest and most complete Mayan text, comprising thirty-nine leaves of pulped tree fibre inscribed on both sides and folded into a book.

It is one of only three historically authenticated codices still in existence. The rest were destroyed by Spanish priests at the insistence of Bishop Diego de Landa of the Archdiocese of Yucatan. As well as recording and predicting equinoxes, solstices and eclipses, the Dresden Codex contains calculations of the phases of Venus and close observations of Mercury. Its final page shows the end of the world: set against a blood-red background, a sky god crowned with snakes pours water over the Earth.

The only known representation of Ixtab, the Goddess of Suicides, is to be found in the Dresden Codex. She appears at the bottom of a page towards the end, hanging by her neck from Heaven. Her eyes are closed, and her mouth hangs open. She has her legs kicked up under her – one of her breasts is exposed, and her hands point down towards the Earth. The black disc on her cheek is a sign of decomposition.

Ixtab was said to lure young men deep into the forest where she seduced them beneath the branches of a Ceiba tree. She is connected with the planet Venus and has close associations with eclipses of both the Sun and Moon – the lunar eclipse having a particularly evil significance for women, especially those carrying unborn children who were exposed to the risk of deformity or death at its height. Most of Ixtab's lovers were never seen again – those who managed to return from the forest would be insane and wish for nothing but a return to her arms.

Known familiarly as 'The Rope Woman', Ixtab's main duty was to accompany the soul of each new suicide to the welcoming shade of *Yaxche*, the Mayan World Tree. She extended the same service to warriors who had fallen in battle, sacrificial victims, women who died during childbirth and members of the priesthood, leading them to the grace and plenty of an afterlife under the protection of the World Tree. Suicide by hanging was regarded by the Maya as a highly honourable way to die, especially for those took their own lives according to Ixtab's favoured method: hanging themselves from the branches of the Ceiba – an act that would instantly gain them a place in Heaven instead of the Underworld.

IV

Reproduced in a popular science book, a detailed copy of two pages from the Dresden Codex haunted the early years of my childhood. Depicting 'the motions of Venus over a 104-year period', it showed angry sky gods at war with the heavens. Outlined in red, green and yellow, a skull-faced deity brandished a heavy club, while others fell into angry argument or 'threatened to kill either the sun or Venus'. A writhing dog pierced by a spear and a fallen warrior gazing upwards in agony marked specific moments of darkness and eclipse. To trace these images back and discover them in the actual Dresden Codex is to be suddenly made aware of the shadowy figures that move behind memory. Even in their cleaned-up and modernised form, the calm indifference of these creatures still has the power to disturb – the product of over a century's stargazing, their eyes remain fixed upon a dimension very different from our own.

V

What circumstances would render suicide by hanging honourable – comparable even to a heroic death? Mayan warriors were encouraged to take their own lives when faced with imminent defeat. Strangulation is closely related to sexual arousal: autoerotic asphyxiation restricts the flow of oxygen-rich blood to the brain, resulting in a state of intense euphoria. 'The hanging gimmick – death in orgasm' as William Burroughs described it in *The Soft Machine*. Erections and ejaculation caused by hanging link the act to fertility and conception: seed is scattered on the earth beneath the dancing feet of a dead man. The objects of Ixtab's care and attention point towards sexual madness and death – more specifically to a physiological astronomy based on human sacrifice.

The accuracy with which the Maya observed the stars was related to the workings of their ceremonial calendar, locating each of their rituals precisely in celestial time. Physiological astronomy aligns human needs and urges with the movements of Heaven. A corresponding shift in perspective is consequently required. Death by hanging remains very different from the other forms of sacrificial killing practiced by the ancient Maya, all of which involved the piercing or dismembering of the victim's body in order to spill the greatest amount of blood. What then made strangulation honourable? It undoubtedly has connections with Mayan methods of agriculture. Slash and burn cultivation requires precise timing: fields of vegetation must be reduced to ash in order to fertilise the exposed soil before the rains come. Any mistake could result in a failed harvest and the very real prospect of starvation. To take one's life during times of drought and famine was therefore a noble thing: an act of self-sacrifice that furthered the life of the community.

More than this, to hang oneself from the branch of a tree can also be seen as a gesture that aligns the victim, hanging between Heaven and Earth, with the *axis mundi.* A special status, spiritual knowledge and power are associated with it. For example, ancient sagas describe the Norse god Odin hanging by his neck from *Yggdrasil,* the World Ash Tree – a name that also translates as 'Ygg's Steed', which is another word for gallows. Physiological astronomy is consequently opposed to one in which geometry and reason establish between them the alignment of human and celestial forces.

The launching by the Soviet Union of a cosmonaut into space was a twentieth-century form of ritual Earth sacrifice: to orbit around the world was to risk danger and death as the capsule passed from night to day and back again in a regular cycle. The first rite of spring took place on April 12 1961 when Yuri Gagarin looked out of his capsule window. 'The sky is dark. The Earth is blue,' he observed. His gaze brought the Earth to life. 'Rays were blazing through the atmosphere of the earth,' Gagarin reported later, 'the horizon became bright orange, gradually passing into all the colors of the rainbow: from light blue to dark blue, to violet and then to black. What an indescribable gamut of colors!' He compared this effect to the spiritual paintings of the visionary scientist Nicholas Roerich.

Hacha'kyum was the divine protector of the Maya. He created the stars by scattering sand in the sky as if it were seed upon a field. At night he walked across the Milky Way, tending his rows. The red letters 'CCCP' painted onto Gagarin's white space helmet, directly above his visor, were added just before take-off. There was a fear that he might be executed as a spy if he were to overshoot the East and come down in the West. A U2 spy plane piloted by Gary Powers had been brought down over Soviet territory the previous spring. Powers died when his helicopter fell out of the sky while he was reporting on bush fires in Santa Barbara County. Gagarin was killed in a plane crash outside Kirzhach in the USSR.

Where else are cosmonauts expected to die, except here on Earth?

SHU

22 May – 21 June

SHU

22 May – 21 June

Also known as:
'Emptiness' or
He Who Rises Up

Body Part:
Lungs, Arms and Shoulders

Areas of Influence:
Shadowing, Division
and Emission

I

Plant a stick upright in the ground, and its shadow will follow the Sun across the sky. A hanging line and a straight edge reveal the heavens. Altars, stone circles and pyramids align themselves with poles and sunrises. A single room with a lamp and a window is a temple to reason. American artist Joseph Cornell referred to the white-walled kitchen of his family home on Utopia Parkway with its stove burning all night as his 'observatory'. Where does the night end and space begin? Witnessing a partial eclipse of the Sun at the age of thirteen, the Danish astronomer Tycho Brahe thought it 'something divine that men could know the motions of the stars.'

'Two things fill the mind with ever new and increasing admiration and awe the more often and steadily we reflect upon them: the starry heavens above me and the moral law within me,' wrote Immanuel Kant, philosopher of the Ideal. Was it as a priest or a farmer that he wrote his *Critique of Pure Reason*? Not everything remains fixed in Heaven. Close observation and measurement reveal that some stars wander across the sky or disappear altogether only to reappear again at certain points in the future, establishing regular patterns over time. The Ancient Greeks called them planets and gave them names. The first star to appear in the sky, Venus, is both the morning and evening star.

Between 1576 and 1580 Tycho Brahe built Uraniborg, his 'City of Heaven' on the small island of Hven in the strait between Denmark and Sweden. Astrologically aligned with the Sun and Jupiter and oriented around the four cardinal points of the compass, it was the first observatory, establishing a new architecture for astronomical research. The largest and most complex instruments were arranged into a theatre of calculations: curves, spheres and grids fashioned in stone, metal and wood allowed Brahe to fix the positions of the planets and record irregularities in the motions of the Moon with a previously unknown accuracy. One device, the Great Equatorial Armillary, was so heavy that it took forty men to move it.

The stars guarantee the fate of humanity – they connect us with virtue and misfortune. In 1572 Brahe observed a new star in the sky, the 'same brightness as Venus', near the constellation of Cassiopeia. It was a supernova. Something in the heavens that should never ever change had just changed. Eighteen months later it disappeared again. It would not return.

II

Astronomer: Today I am answering questions about the sun. Ask me anything you like.
Question: Where is it?
Astronomer: ...

A field of sunflowers turning their heads in summer will always find the Sun. Still the eye catches the flash of a shopping bag on the street with *'Mais il est où, le soleil?'* printed on it.

III

Our eyes occupy themselves in the darkness, seeing nothing and yet still finding things to see. Astronomy becomes a mechanised ceremony designed to generate meaning. The Ancient Egyptians were among the first to make a close study of the stars, arranging them into patterns reflecting the theology behind their rituals. The dawn became a daily re-enactment of the world's creation: the first light on the horizon marking the separation of earth and sky. They identified the Pole Star as 'the great city': architectural forms establishing the boundaries and shape of space. Shrines and temples form their own cosmology of symbols and representations. 'The shrine of the god, for instance, was "the Horizon", the land of glorious light beyond the dawn horizon where the gods dwelt,' according to one historical account. 'The temple was an image of the universe as it now exists and, at the same time, the land on which it stood was the Primeval Mound which arose from the waters of the Primordial Ocean at creation.'

The waters of the Primordial Ocean stretched out in all directions. An endless depth without a surface, its limitless darkness extended to the very edge of human knowledge. It assumed no shape or substance: undivided and undisturbed by dimensions, direction or distance. While this world was still flooded with silence, the Primeval Mound emerged, breaking space open – and the great city of Heliopolis arose. Dedicated to the worship of Atum, the first complete being, and later to that of Ra, high god of the Sun, Heliopolis was the greatest religious centre in the whole of Ancient Egypt. It represents the imposition of spatial qualities upon space: it is the bodying forth of creation on the physical plane of existence. Atum, who comes before all things, is often referred to as residing in Heliopolis. An early sarcophagus inscription has Atum speaking of his existence 'before Heliopolis had been founded that I might be there', when he was still not a part of this world.

Atum, who comes before the appearance of the sky and the earth, the appearance of men, the birth of the gods and the appearance of death, drifted as a spirit through the Primordial Ocean: unformed and inert and without meaning. Finally appearing in visible form, he becomes the scarab-faced god Khepri or 'He Who Is Coming Into Being'. Atum is also, according to the sarcophagus text, 'that Great He/She' who is neither man nor woman nor both. The primary and self-created being, Atum 'proceeded to masturbate with himself in Heliopolis; he put his penis in his hand that he might obtain pleasure of emission thereby.' A brother and sister were born of this divine ejaculation – Shu and Tefnut, who were the original pair and marked the first division of reality. 'I am that space which came about in the waters,' Shu declares in the same sarcophagus text. 'I came into being in them.' Claiming to be the embodiment of space, Shu connects the unity of Atum, the first progenitor, with the teeming multitudes of creation; and his sister Tefnut represents the introduction of order into the chaos of the universe.

Just as the structures of Heliopolis frame the presence of Atum, so in the unfolding of physical space Atum gives birth to Shu and Tefnut in the same city. Tefnut is also associated with the tongue and human speech, while the name 'Shu', which means 'emptiness', is closely associated with the word for 'to spit'. The divine ejaculate becomes closely associated with the mouth spitting. 'This was the manner of your engendering,' Shu says to Atum. 'You conceived with your mouth and you gave birth from your hand in the pleasure of emission.' Spitting in turn equates with the forming and uttering of words and, through Shu and Tefnut, with the words' connection to the breath of life. Human speech, the act of naming, which is also that of dividing and separating, is consequently linked with masturbation and the shooting of semen from the penis.

More than any other god, Shu is ashamed of his own becoming. Creation is rendered unspeakable through him. The Sun's rising is obscene. Speaking and naming, dividing and ordering are, through his example, linked with creation and therefore exposed as indecent acts. The uttering mouth is a shameful union between air and space, breath and tongue. Shu and Tefnut consummate the relationship by coming together to produce another brother and sister pairing: Nūt, the sky, and Geb, the earth. Incest and separation are mythically linked through this union.

The Ancient Egyptians believed that Shu created space and carries it above his head with empty arms. Shu separated his son Geb from sexual union with his daughter Nūt and 'so loved her' that he lifted Nūt up into the sky, where she gave birth to the stars. The Milky Way becomes a spurt of semen across the night sky; while Nūt transforms the night sky into flesh, taking her children back into her body to travel through the darkness from dusk until dawn.

'I shall join the whole earth to you in every place, O High above the earth!' Shu declares to his daughter

Nūt, according to one pyramid inscription. 'You are supported upon your father Shu.' Geb falls stricken and abandoned to the ground, while wind spirits support Shu's arms and shoulders as he raises Nūt up. 'I am he who joins and separates,' he announces. The joining and separation of Geb and Nūt is enacted with the closing of the coffin in the burial chamber: the sarcophagus and its lid coming together like the earth meeting the sky.

IV

In Mesopotamia over three thousand years ago the occurrence of a spring equinox is noted on a clay tablet. Twenty-one centuries later the administrator of an Egyptian temple near Thebes witnesses an eclipse of the Sun that occurs on 10 March at midday: a fragment of pottery still bears his account of the event. His name was Petros.

Four hundred years before our calendars started counting forwards from zero to the future, at a time when the kings of Persia still ruled an empire, the wanderings of Mercury and Mars were inscribed upon clay tablets. And a thousand years before that on the West Bank of the Nile the conjunction of Jupiter, Saturn, Mercury and Venus with Sirius was depicted on the ceiling of a tomb in the Theban Necropolis. Separated from this planetary grouping, Mars was represented by an empty boat: appearing to move backwards from its usual course in the night sky every two years or so, due to a discrepancy in its orbit relative to the Earth, the planet was consequently thought by the Egyptians to be false and misleading.

V

The bigger astronomical instruments became, the more accurate they were to the naked eye. Between 1724 and 1734 at Jaipur and New Delhi the Maharaja Sawai Jai Singh raised monuments to the heavens. Great arcs, and diagonals, extended staircases, planes and wells provided fixed angles by which to check stars' positions and correct the Indian calendar. The building work remained uncompleted, however: the giant sundial that cast a shadow onto calibrated masonry fell into disuse. The ruins of these celestial palaces became ever more otherworldly. Their stately geometry was resurrected in the fantasy towers and minarets of Luna Park, one of Coney Island's star attractions at the start of the twentieth century. Once beyond the colourful half moons adorning its entrance you stepped onto a lunar landing site inhabited by Moon Maidens. Wander out into this fantasy world and hear the people screaming with joy, speeding down the steeply inclined runway of the 'Shoot-the-Chutes' and out onto a wide lagoon. 'You see, this being the Moon, it is always changing,' its founder boasted. 'A stationary Luna Park would be an anomaly.'

Archaeologists using commercial satellites have discovered a huge structure buried beneath the sand close to the centre of the ancient city of Petra in Jordan. WorldView-1 and WorldView-2 helped to map the faint outlines of a ceremonial site that has not been visible from ground level for many centuries. Over two thousand years old, its existence has barely been a suspicion: satellite images indicate a dusty floor plan of the phantom building imprinted in the grey sand, comprising a vast platform, flagstones, columns and a huge staircase. Through its network of satellites outer space functions as a dark mirror, throwing back images of the life captured below. Operating 300 miles above sea level and orbiting the Earth every 1.6 hours, WorldView-1 can collect up 290,000 square miles of imagery in any 24-hour period. WorldView-2 has an advanced on-board system that 'can capture pan-sharpened, multispectral images' from almost 500 miles out, photographing over 350,000 square miles daily.

Soviet cosmonaut Valentina Tereshkova, the first woman to travel into space, has described how her parachute opened just as she was approaching a lake on her return to Earth; 'a strong wind carried me over the mirror-like water to the opposite shore,' she recalled, as the dark sphere of her Vostock capsule glided over its own silent reflection in the early morning sunlight. It remains a moment of perfect peace: an image caught floating towards its earthly home.

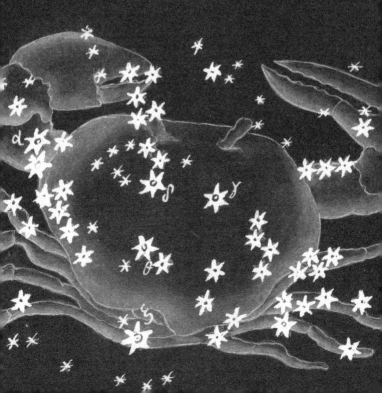

LEVANIA

22 June – 22 July

LEVANIA

22 June – 22 July

Also known as:
Selene or
'An Island in the Moon'

Body Part:
Chest, Breasts and Stomach

Areas of Influence:
Reflection, Isolation
and Conjuring

I

Childhood experience reveals that it is impossible to examine the craters of the Moon through a pair of toy binoculars from the back seat of a moving family car at night. However much you squint and try to focus, the scratched plastic lenses reveal no more than a yellow blur lacking any detail and all awe. The city spins by, its streets an unnoticed smudge of light. I had wanted to see the Moon's mountains and valleys and the long shadows that they threw in books and magazine illustrations. The binoculars were Made in Hong Kong and bought at Woolworths – they may also have spent more than a few nights left out on the back lawn. 'The island of Levania is located fifty thousand German miles high up in the sky,' a daemon reveals in Johannes Kepler's *Somnium*, a short astronomical text published after his death in 1630. Recounting a dream experienced after falling asleep while 'contemplating the stars and the moon', Kepler claims that he 'seemed to be reading through a book that I had brought from the market' that recounted the adventures of an Icelandic youth named Duracotus. The son of a witch, Duracotus studies astronomy with Tycho Brahe at Uraniborg, his island observatory off the Danish coast. The Moon, which the daemon knows as 'Levania', is transformed into an offshore island orbiting the Earth. 'The route to get to there from here, or back to this Earth, is rarely open,' the daemon reveals. 'When it is open, it is easy for our kind, at least, to travel. But transporting humans is truly difficult, and risks the greatest dangers to life.'

Reflecting the Sun's light, the Moon misleads us, contrarily moving around the Earth while everything else in the Solar System chooses not to – the only piece of celestial mechanics that connects the geocentric with the heliocentric systems, which cannot occupy the same universe as each other. Meanwhile the Moon remains the Moon – pure as a blank screen.

The Moon struggles with temptation, which is another form of enchantment, looking down on lovers and painters captured by its borrowed light. Every poet and every astronomer who has ever lived has been transported by moonlight. 'Autumn Moon over Miho' from Utagawa Kunisada's series *Eight Views of Beauties* shows a woman reading by the light of the Moon, her breasts exposed. The elegant curve of her body connects the late evening sky with the wooden verandah upon which she is standing. The Moon is full and low in the sky – standing out against washes of blue ink, little more than a blank circular space on the printed page. A tiny dark blot, like some distant world, appears to be circumscribing its lower hemisphere. Behind the woman and unnoticed by her as she inclines her head over her letter, a second Full Moon is shown, rising above the autumn treeline, on a round disc of paper. Captured between these two bright Moons on a printed page, she continues with her reading.

When did the Moon stop being a disc and become a sphere? Even as it recounts a tale told in a book perused by a dreamer, Kepler's *Somnium* concerns itself primarily with describing the effects that the Earth, Moon and Sun have upon each other as physical bodies moving through space. Imagine looking at the Moon every night and not understanding what it was – just a big glowing circle offering a monthly reminder of creation, waxing and waning over the succession of nights. *Moon is the Oldest TV* by Nam June Paik recreates the phases of the Moon as a series of magnetically distorted fields on a line of black and white monitors in a silent, darkened room: restoring the early mystery of the lunar cycle at a time when I was gazing at it through my cheap plastic binoculars. '*TV will dominate 50 years and... will gone...*'

Paik writes in a letter to John Cage. '*What comes next? How white, black, yellow people spend their leisure time in the earth between their vacation in the moon???*'

II

A disc, according to Euclid, is the region on a plane bounded by a circle, which is also the curve traced by a point moving in such a manner that the distance it keeps from another given point remains constant. The circle is divine: the definition of eternity, without beginning or end – it also forms one section of a sphere. 'See now, I will rise,' says the Sphere, moving upwards through two dimensions in Edwin A. Abbott's *Flatland*, 'and the effect upon your eye will be that my Circle will become smaller till it dwindles to a point and finally vanishes.' A sphere is the set of all possible points existing at the same distance from another given point in three dimensions. In early depictions, the Sun and Moon appear as flat discs moving along a semi-circular path that leads from one side of the sky to the other: their 'given point' is the Earth, which is constant but not eternal or perfect.

Featureless, regular and complete, the faces of the Sun and the Moon are readings in two dimensions of three-dimensional objects. They represent the collapsed perspectives and the blank horizons of the untrained senses. Space is an undeveloped image to the naked eye. Distance is projected before it is ever measured.

We see faces because we put them there. 'The human head is formed upon the model of the celestial spheres; it attracts and it radiates, and this it is which first forms and manifests in the conception of a child. Hence the head is subject in an absolute manner to astral influence,' according to Éliphas Lévi in *The Dogma of High Magic*. The sphere only becomes real to us by analogy: the roundness of the human skull and the space that it occupies are its true model.

And what form does knowledge take? How is certainty maintained? 'The senses roll themselves in fear, and the flat earth becomes a ball,' observes William Blake in the sixteenth stanza of 'The Mental Traveller'. Kepler's daemon uses similar terms to describe how a human copes physically with the journey to Levania. 'When we reach the open sky,' he reveals, 'we remove our hands from his body so that he balls himself up like a spider, which we transport almost by our will alone, so that finally the mass of the body falls towards the intended destination of its own accord.' The body responds to space by protecting itself, head pulled down instinctively towards stomach and chest, the back rounded into a sphere.

III

In order to explain the motion of the twin discs and the other points of light across the sky the Greco-Egyptian astronomer Claudius Ptolemy developed the theory that they are all embedded in ten separate crystalline spheres, one contained within the other and with our world as their common centre. This first and inner sphere is material, comprising earth, water, fire and air, and is therefore imperfect and impermanent. Next come the seven spheres of the planets, including the Sun and the Moon, followed by the sphere containing the fixed stars and the circle of the Zodiac. The tenth and outermost sphere is both Heaven and Prime Mover, according to which everything moves in its perfect and eternal course from east to west. Drawing upon eight hundred years of observations and measurements, Ptolemy devised a system that could predict these movements with greater accuracy than ever before.

'Brahe and his students watched the moon and constellations all night with marvelous instruments,' Duracotus reveals. 'These activities reminded me of my mother – in fact she had a constant habit of talking to the moon.' Stars that appear and disappear again did not fit Ptolemy's predictive cosmological tables – nor did the Great Comet that passed close to the Earth in 1577. Comparing observations from Copenhagen and Prague, Tycho Brahe determined that comets existed somewhere beyond the Moon. He even published a pamphlet denouncing the Great Comet as evil, 'which was revealed by its pallid appearance and unclearly shining color like the star Saturn.' Kepler had studied with Brahe on his island – just as Kepler's mother had been publicly accused on two separate occasions of being a witch.

IV

Over the history of astronomy, the object becomes a moment of perception: even the most celestial body is no longer eternal but is forced to exist in time, shaping and reshaping itself on the edge of our senses. There is the Sun of 'mathematical serenity and spiritual elevation' to which our eyes have long been adapted. There is also the Sun that consumes human sight – the one that shines until it is black, burning itself into the back of the eye as a dark quavering dot. In his *Philosophia Reformata*, first published in 1622, Johann Daniel Mylius uses a Black Sun to represent the stages of putrefaction in the alchemical process. The body, he notes, 'is also dissolved by its own heat and humidity' in a material form of madness. Finally there is the Sun projected through a pinhole onto a sheet of white paper – a miniature solar image, inverted yet clear. An expanded version of this optical effect was used in 1544 to project the solar eclipse of January 24 upon a darkened chamber wall.

Light rays entering the pinhole and hitting the wall are virtually parallel and therefore in focus. *Camera obscura* images, projected into the dark spaces of the technical mind, are dreams reproduced at a distance, their colour and perspective preserved. NASA's New Worlds Imager, a giant interstellar projector, will align a star through a nine-metre hole at the centre of a giant opaque plane with a probe positioned tens of thousands of miles behind it. This allows the close observation of worlds orbiting that star without the distraction of scattered light and the optical imperfections contained within lenses: capturing oceans, continents, polar caps and clouds from a hundred light-years away.

In 1646 the Jesuit scholar Athanasius Kircher publishes *The Great Art of Light and Shadow*, his treatise on the projection of images in a darkened room. The frontispiece shows beams of light emanating from God and directed by mirrors, pinholes and tubes across an ordered landscape. In 1655 *The Anatomy of the Sun* documents Kircher's solar observations. The Sun appears to consume itself with its own corruption, its face pitted with mountains, releasing dark plumes of matter. Kircher also studies Egyptian hieroglyphics, which he considers to be the visual representation of the language God taught to Adam and Eve.

V

In 1543 Nicolaus Copernicus published his treatise *On the Revolutions of the Celestial Orbs*, describing a perfectly geometrical system with the six known planets circling the Sun – the Earth being alone in having its Moon. Kepler's *Somnium*, for all its dreamlike detail, describes a heliocentric arrangement between these three bodies from dawn to dusk and from Levania's equator to its poles. It was written in 1609, the same year Galileo trained his first 'optical canon' on the heavens 'to the infinite amazement of all'. To the progressive mind, science is always limited by the age. Galileo was above all an illustrator and a writer. Lenses and printing clarify each other. Extensions of the telescope, books exist as astronomical instruments in their own right: each has its own centre of gravity into which we fall. Published in 1610 and sounding like a Papal decree, Galileo's *Sidereus Nuncius*, or *Starry Messenger*, reveals that the Moon is not a perfect sphere but a separate and uneven world of mountains, ridges, valleys and shadows. His telescope reveals that Jupiter has four large orbiting stars, meaning that our Moon is not alone in Heaven.

The *Selenographia* produced in 1647 by Johannes Hevelius of Gdansk was the first treatise devoted to the Moon. Named after Selene, lunar goddess of the Ancient Greeks, the book presents its reader with every possible phase of the Moon recorded over a four-year period. Its 110 plates, engraved by Hevelius himself, use crescents of blank paper to represent dark phases of the Moon, two empty circles marking its maximum extent, while the illuminated lunar surface gradually appears in widening slivers. The book's frontispiece aligns Hevelius with the first-century astronomer Ibn al-Haytham 'by reason' and Galileo 'by the senses'. What the eye sees for itself becomes more important than the object's geometrical orientation in space. Hevelius presents his Moon as it would have appeared through the lens of a telescope – upside down.

GALILEO

23 July – 22 August

GALILEO

23 July – 22 August

Also known as:
Panzerfaust or
The Stargazer

Body Part:
Heart and Upper Back

Areas of Influence:
Reflection, Isolation
and Conjuring

I

'Scarcely have the immortal graces of your soul begun to shine forth than bright stars offer themselves in the heavens which, like tongues, will speak of and celebrate your most excellent virtues of all time,' Galileo writes in a letter seeking patronage from his former student Cosimo de Medici, now the Duke of Tuscany. The astronomer proposes that the four moons he has recently discovered orbiting Jupiter should be named the 'Medicean Stars' after Cosimo's family, being a powerful house of bankers, merchants and popes. 'And thus the Sun, as if seated on a kingly throne, governs the family of planets that wheel around it,' declares Copernicus of the royal ordering of the Solar System, evoking the 'thrice great' alchemist Hermes Trismegistus.

Names suggested for the Galilean moons include *Principharus, Victripharus, Cosmipharus and Fernipharus*, all of which pay obsequious homage to the Medicis. Meanwhile Johannes Kepler suggests naming them *Io, Europa, Ganymede* and *Callisto* after Jupiter's lovers, which Galileo flatly refuses to do – ever. Instead he elects to refer to the moons simply as numbers, starting with the one nearest to Jupiter itself and then counting outwards. Galileo also makes a point of not mentioning Kepler's name anywhere in his *Sidereus Nuncius*.

Kepler uses Tycho Brahe's detailed astronomical tables when developing his heliocentric cosmology. 'I confess that when Tycho died,' he later reveals, 'I quickly took advantage of the absence, or lack of circumspection, of the heirs, by taking the observations under my care, or perhaps usurping them.' The planets become bodies following elliptical paths around a physical Sun that exerts its influence directly upon them. Kepler studied theology, wrote about daemons on the Moon and stole from a dead man. Brahe lost a chunk of his nose in a duel. You still see their names and likenesses carved into museum walls everywhere.

II

In 1599 the Franciscan monk and astrologer Tommaso Campanella is imprisoned and tortured for using magic and for attempting to establish a utopian society.

In 1616 Galileo Galilei is reprimanded by the Catholic Church for suggesting that the heliocentric cosmology proposed by Copernicus can be proved merely by observation of the Moon and stars.

In 1622 Campanella writes an *Apologia pro Galileo, mathematico fiorentino*, defending him from his prison cell. Also supporting Galileo's position is his patron and friend Cardinal Maffeo Barberini, who becomes Urban VIII in 1623.

In 1626 Urban VIII orders Campanella's release from prison and brings him to Rome. Plagued by ill health, political intrigues and persistent rumours of his impending death, Urban VIII asks Campanella to use his astrological knowledge to protect him from the fatal influence of a lunar eclipse predicted to take place in 1628. Did Galileo's observations make the Moon more real for Urban VIII? Eclipses, like comets, bring tragedy with them. The Pope's enemies gleefully claim that this one will kill him. Campanella, however, knows how to dispel the harmful effects of a 'darkened menacing sky' and 'what odours and tastes, colours, temperature, air, water, wine, clothes, conversations, music, sky and stars are to be used for breathing in the Spirit of the World, which is implanted and inserted in its individual parts and is diffused through the whole of it, and under what constellations.'

A suite of rooms in the Lateran Palace is set aside for a magic ceremony timed to take place during the hours of the eclipse. The walls of a designated chamber are covered in white silk and decorated with foliage. Two lamps are lit to represent the Sun and Moon respectively, then five torches for the known planets. Other lights stand in for the Zodiac. The palace apartment's doors and windows are sealed against the 'noxious seeds' carried in air polluted by an eclipse; and the room is purified by the 'sprinkling of essences and distilled waters' and the burning of aromatic woods such as laurel, myrtle, rosemary and cypress. All of those involved in the ritual are selected because they are astrologically impervious to the darkening of the Moon. They too are dressed in white. The 'playing of soothing music' connected with Jupiter and Venus is an important part of the ceremony, as are 'enticements' in the form of gems, plants, perfumes and colours capable of attracting the benign influence of the stars and countering the malignant ones.

In 1629 Campanella's short treatise *De siderali fato vitando*, 'How To Avoid the Fate Dictated by the Stars', is published without Papal approval or the author's consent. By exposing details of the ritual conducted at the Lateran Palace, the Pope's enemies have set out to embarrass and discredit him. Plans to make Campanella a member of the Holy Office are hastily reversed; and he is charged instead with heresy, while Urban VIII denounces those who use astrology for the purposes of divination. Don Orazio Morandi, abbot of the monastery of Santa Prassede, is imprisoned for predicting that the Pope will die during the solar eclipse due to take place in 1630. Meanwhile Campanella writes an apology for astrological prognostication, ingeniously aligning it with Church doctrine: 'by asserting that astral fate cannot be avoided,' he argues, 'one subjects free will to the stars.' Morandi subsequently dies in prison amid rumours that he was poisoned on the Pope's orders.

In 1632 Galileo infuriates Urban VIII by publicly mocking him in his *Dialogue on the Two Chief World Systems* and is subsequently forced to abjure the Copernican system: his book is suppressed and he is placed under house arrest. How else to avert scandal in an age when eclipses are both routinely predicted and guarded against through magical rituals? Campanella eventually finds a home in the French court, where his last published work is a poem celebrating the birth of the Sun King: Louis XIV.

III

Lenses magnify planets and kings. A bee is a device for transforming sunlight into sugar. Pope Urban VIII elevated his brothers and nephews to high offices both in the Church and in Rome, allowing the Barberini to amass a huge fortune. *What the Barbarians did not succeed in doing to Rome*, runs one joke from the time, *the Barberini achieved.* The bee was their family emblem – swarming to suck the honey from the age.

'A very barren and thin hint of the Plot I had from the Italian,' writes Aphra Behn of her play *The Emperor of the Moon*, 'and which, even as it was, was acted in France eighty odd times without intermission.' First performed in 1687, it recounts how two Neapolitan gallants trick an addled astronomer, Doctor Baliardo, into thinking that they are lunar dignitaries arrived on Earth to court his jealously guarded daughter and niece. Everyone in *The Emperor of the Moon* agrees that the doctor's wits have been turned by his obsession with books and charts and celestial instruments. One member of Baliardo's household describes how 'he calls up all his little Devils with horrid Names, his *Microscope*, his *Horoscope*, his *Telescope*, and all his *Scopes.*' Furthermore, Baliardo believes that the Moon is inhabited and uses a 'Telescope twenty (or more) Foot long' to spy on what he believes to be its King in his private chambers.

At the play's conclusion, set in an old gallery 'richly adorn'd with Scenes and Lights', Baliardo is amazed to witness Kepler and Galileo come down from the sky 'in Chariots, with Perspectives in their Hands, as viewing the Machine of the Zodiack'. Accompanied by 'the Musick of the Spheres', they greet him as ambassadors on behalf of the Emperor of the Moon and 'the Prince of Thunderland'. In turn, Baliardo hails them as bards and philosophers, just as Campanella publicly defended Galileo as a mathematician rather than an astronomer. Then Kepler urges Baliardo to 'look up and see the Orbal World descending; observe the Zodiack, Sir, with her twelve Signs.' The Zodiac descends, 'a Symphony playing all the while', after which the twelve Signs sing and perform a dance to Baliardo's wonder and delight.

Finally Kepler reveals that he is really Baliardo's personal physician in disguise and explains that 'It was not in the Power of Herbs or Minerals, Of Reason, common Sense, and right Religion, To draw you from an Error that unmann'd you.' Instead this elaborate ceremony of costumes, music and lights, involving the procession of stars and planets, has been staged in order to cure him of his lunar folly. As the farce draws to a close, Dr Baliardo welcomes the two gallants into his family and then orders that his books be burned – 'let my Study Blaze. Burn all to Ashes, and be sure the Wind Scatter the vile Contageous Monstrous Lyes,' he commands before declaring that 'there's nothing in Philosophy.'

IV

If the darkness during an eclipse persists, Campanella suggests in *De siderali fato vitando*, it may also be necessary to secure further protection by constructing an artificial sky.

George Herriman's Krazy Kat impassively watches as the Moon comes crashing down into the ocean's dark and distant horizon – a feat possible only in ritual magic and comic strips. 'I just knew that "Moon" would fall down some day,' he says to nobody in particular.

V

Found the main entrance, where a uniformed guard directed us to reception. We handed over our passports at the desk and were photographed by ceiling-mounted cameras, cool and remote, angled down above our heads. We were immediately issued with passes that had our pictures and a barcode printed on them – they were collected from us again when we finished the tour.

There was a moment of recognition: a flash of memory from movies, television news and magazine covers as we were guided down one of the main corridors, broad and white and open. We looked in through sealed glass partitions at clean rooms and large assembly areas fitted with massive flat tables designed to support large and heavy satellites. Occasionally we would catch a glimpse of human hands at work – figures in white protective suits leaning over wide sheets of carbon fibre panelling.

We saw the thermal shield for a solar probe designed to keep extreme levels of heat away from delicate instruments that can only function at room temperature – very few people will ever see the back of this shield, we were told.

We saw quartz manufacturing laboratories and sewing workshops, all connected by the same white corridor. Here were the glittering components of satellites – the stitching of golden fabrics and the growing of crystals beneath bright lights. Outside of this white corridor the Earth seemed grey and abandoned: fading modern architecture of the 1950s in red brick and painted cement.

On 26 April 1945, dressed in the uniform of a colonel in the Soviet Army, the Russian rocket engineer Boris Chertov drove up to a nondescript brick building located in Adlershof, which is now called the new city of Science, Technology and Media outside Berlin. The gates had been blown off their hinges during the Allied advance, and the body of a dead soldier was still lying inside the entrance – shot through the back and chest. Everything inside the secret research facility was still intact – the equipment, the precision tools, the files of test results and blueprints. The gleaming white laboratories with their drafting tables, vibrating benches, calibrators and oscilloscopes had been left deserted. Steel safes filled with classified documents stood abandoned and untouched. At that precise moment Nazi rocket scientist Wernher von Braun was somewhere outside Peenemünde, nursing a broken arm and planning his surrender to the Americans. Moving down the white corridors Chertov thought it was a shame to break open the shining sets of steel doors. 'I am stepping over the body of a young *panzerfaust* operator that has not yet been cleared away,' he would later write in his diary. 'I am on my way to open the next safe.'

The white corridor is not a dream. It has existed for years: a shifting location taking shape in our memories from moment to moment. A procession of machines, engines and bodies has passed down this corridor, which stretches from the past deep into the future. Astronauts, already sealed into their helmets and breathing their own air, have followed it into a universe bounded by the ducts and pipes, channels and chambers that honeycomb space.

URANIA

23 August – 23 September

URANIA

23 August – 23 September

Also known as:
The Astronomer's Muse or
The Highest Star

Body Part:
Abdomen and Digestive system

Areas of Influence:
Hygiene and Efficiency,
Command and Control

I

Wernher von Braun's mother was Baroness Melitta Cécile von Quistorp. Descended from Swedish nobility, she shares with him her love of astronomy and music. It is upon her recommendation that von Braun chose Peenemünde, located at the sandy mouth of a river on an island close to the Baltic Coast, to build an Uraniborg for the twentieth century. Greatly extending the reach of Tycho Brahe's City of Heaven, this new research centre is devoted, under von Braun's direction, to charting outer space as the complex interplay of distance, mass and velocity. Wernher von Braun has plans to travel to the Moon some day. Meanwhile Peenemünde is the place where they will design, develop and build rockets for the Nazis.

Cinema is rocketry for the eye – particularly when it comes to the depiction of distance, mass and velocity. In 'The Astronomer's Dream', a short film from 1898, Georges Méliès uses movie camera trickery to show the Moon eating a telescope. In 'A Trip to the Moon', made in 1902, the astronomers wear tall conical hats and long robes that make them look like wizards. They pose and gesticulate in front of a giant painted canvas showing a telescope, a globe and a large armillary sphere. The astronomers transform their telescopes into stools so that they can sit down. They plan to take a trip to the Moon.

On von Braun's island rocket base dinner is an elegant formal occasion: lines of waiters in black suits, white shirts and bow ties serve at tables laid with crisp linen and silverware. The four-star hotel attached to the research facility has a well-stocked wine cellar containing vintages looted from France by the victorious German army. Von Braun plays cello in a string quartet with three other rocket scientists at Peenemünde – their repertoire favours works by Mozart, Haydn and Schubert.

Cinema transforms the Moon from a lighting effect in the background of paintings to a star. Méliès's Moon has a man's face on it. The rocket from Earth hits him right in the eye. Jump cuts, fades and false perspectives collapse space to a moment of impact.

After Wernher von Braun comes to America all evidence of his connections with the Nazi party and his promotions within the SS is either destroyed or sealed in a vault until 1984 – seven years after his death. One photograph still in existence records a visit made by *Reichsführer-SS* Heinrich Himmler to Peenemünde in April 1943. *Sturmbannführer* Von Braun can briefly be glimpsed in his black death's head uniform, his features thrown into partial eclipse by Himmler's round face moving through the shot like an unsmiling moon.

II

Urania, the muse of astronomy, dresses in a cloak embroidered with stars and keeps her eyes focused on the heavens. Astronomy was classified as an art, alongside Music, Geometry and Mathematics, in the mediaeval *quadrivium* of subjects that could be taught to advanced minds. Each provides a path to the closer understanding of the Divine Will. Painting, sculpture and poetry were regarded merely as the decoration of eternal truth, which also marked the triumph of allegory over representation. A fifteenth-century engraving for *The Practice of Music* shows a three-headed serpent connecting Heaven with the Underworld – the Earth, the four elements and the seven known planets are arranged down the chord of its body, separated by tones and semitones. Urania is positioned at the top of the chord, in close proximity to the curl in the serpent's tale, linked with the starry celestial orb and placed immediately below the throne of Apollo, who sits above all of Creation holding a lute.

Pythagoras had already worked mathematics, music and the contemplation of the heavens into a coherent philosophical proposition. His unheard 'Harmony of the Spheres' was to be demonstrated as a spiritual truth through observation and reason. The intervals between notes in a harmonious progression when played on a single chord represent a simple numerical progression. Frequency and length are connected through pitch. The principal celestial bodies move in regular patterns through the sky establishing a melody whose oscillations can only be felt by creatures living below on the Earth.

Johannes Kepler and the alchemist Robert Fludd were both students of Pythagoras and integrated the divine harmony of the cosmos into their theories, even though they viciously attacked each other in public. The two had more in common than either might have admitted. Alchemy was sometimes referred to as 'terrestrial astronomy' precisely because of the harmonic correspondences between Earth and Heaven. The cosmos is transformed into the sphere of human creativity and maintains itself as such by containing our world within it. The greater macrocosm is reflected as a well-ordered whole within the individual microcosm.

The Earth, in other words, is always at the centre of our Universe, whether we like it or not.

The link between the two is based upon conceptual similarities. 'As above so below', the ancient alchemists believed: human experience becomes mirrored in the arrangement of planets and stars. The seventeenth-century physicist Evangelista Torricelli claimed that the creatures of Earth lived their lives 'submerged at the bottom of a vast ocean of elemental air', the weight, pressure and friction of our atmosphere extending up as far as the stars. Seen from the cosmic perspective, the Russian scientist Vladimir Vernadsky proposed that all planetary development – from the geosphere of mineral phenomena to the biosphere of plants, insects and animals – is shaped and bounded by outer space. Humanity, according to his theory, was always destined to occupy the noosphere, which marks an evolutionary advance in consciousness, as cognitive creatures.

With Galileo's discoveries, astronomy ceases to be an art. 'Descend from Heav'n Urania, by that name, If rightly thou art call'd,' John Milton writes in Book 7 of *Paradise Lost* – although he is very careful in the phrasing of his invocation to make it clear that it was 'The meaning not the Name I call'. The Urania cinema, with the signs of the Zodiac circling its vast cupola and the Last Judgment painted on its walls, was to be one of the centrepieces planned by Albert Speer for the rebuilding of Berlin after the Nazi victory in 1950.

III

Saturn makes space look weird: tiny cartoon characters dancing and chasing along its rings like a multilane racetrack. Everyone from Fleischer to Disney to Osamu Tezuka has used that image to depict the highest and strangest star in the heavens. Warned that his offspring would dethrone him one day, Saturn devoured all but one of his children. It was Jupiter, the last and youngest, that evaded and eventually killed him. Old and grey-haired, bearing a scythe and eating his babies alive, Saturn was never going to make it in Hollywood the way the other gods did. Mars, Venus, Neptune and Mercury all had promising film careers ahead of them. Jupiter, in particular, was destined for the big screen. Movie stars are always something special.

Having arrived at their destination, the astronomers in Georges Méliès' 'A Trip to the Moon' bed down for the night among the craters. Celestial forms appear while they sleep. A cardboard comet spins by; the asterism Ursa Major, made up of smiling women's faces, fades in and out of the darkness. Finally Urania and Selene appear as female beauties draped in Greek folds. Saturn is pictured right next to them: an angry old man with long white hair and a beard leaning out of a large hole in the main body of his planet. He shakes his fist at the sleeping lunar explorers and helps Selene, who sits in the hollow crescent of a waning moon, to make it snow upon the sleepers. Urania, Selene and Saturn slowly fade into the thickening whiteness, never to be seen again.

In July 1609 Galileo shares a secret with his patron: 'the star of Saturn is not a single star, but is a composite of three, which almost touch each other, never change or move relative to each other, and are arranged in a row along the Zodiac, the middle one being three times larger than the lateral ones, and they are situated in this form: oOo.' Galileo later circulates a 37-letter Latin anagram concealing the statement: *I have observed the highest planet tri-form*. In his mythological study *Saturn Devouring His Son*, painted in 1636, Rubens depicts the highest planet as a configuration of three stars: one large and two small six-point snowflakes in a darkening sky. Saturn himself is stooped and grey haired, concentrating so closely on tearing the flesh from his baby's chest with his teeth that he cannot hear its screams.

Lenses and words define form: both function on the outer edge of clarity. Saturn rules over the madness of fertility and decay and is associated with the Black Sun: alchemic symbol for corruption. He lives by digesting the flesh of his own flesh. Saturn also castrated his father Uranus – the spilt blood bringing forth Venus from the sea.

Stars and planets had forever been points of light, but Saturn was something different. Some astronomers claimed that Saturn was oval in shape. Galileo observed that the two lateral bodies could appear and disappear at regular intervals – as if dim and distant Saturn were eating its celestial offspring once more. Hevelius suggested that the planet was an ellipsoid with crescents, appearing as handles on either side. Using a more powerful telescope Christiaan Huygens of Holland discovered Titan, Saturn's largest moon, and determined that the 'triple-bodied star' was actually encircled by a ring. His *Systema Saturnium*, published in 1659, details its changing appearance over time: a foldout chart bound into the book shows thirteen separate permutations. Images of Saturn, based on Huygens' detailed observations, are so emphatically clear that the ink has soaked into the pages, leaving phantom ringed planets looming behind the words.

IV

The rings of Saturn are composed of water-ice and rock. They are alphabetically designated in the order of their discovery, which does not reflect in any way their relative proximity to the main planet – hence D, C, B, A, F, G, E and that is not all. There is also the Cassini Division, the Roche Division, the Janus/Epimetheus Ring, the Methone Ring Arc, the Athene Arc, the Pallene Ring and the Phoebe Ring. As of this writing Saturn has 62 detected moons, nine of which remain unnamed.

V

An illuminated billboard flashes the new advertising slogan for a fast food chicken dinner: 'Leave Space For It'. How is that even possible – abandoning the cosmos for fried chicken, however great it tastes? Urania holds out a celestial globe, pointing at it firmly with a small staff. By the late sixteenth century she had become known as 'the Christian muse'.

Cosmonaut Aleksandr Lazutkin was 'desperate to fly like a bird' and believed that this could 'only be done in space.' Vladimir Vernadsky considered the human condition to be a purely transitional one. Humanity, he argued, was about to reach an advanced level in the organisation of its physiological and biological activity, to become a creature uniquely adapted to inhabit the noosphere. Space, like a movie camera, has the power to make everyone a star. Rockets will help to do the rest.

Aleksandr Lazutkin has noted how 'a light breakfast and enemas were helpful in building morale' on the morning he took off into microgravity to join the crew aboard the Mir space station. Surgical spirit is also used to disinfect the cosmonauts' bodies to prevent them from carrying microbes up into the sealed human containers orbiting high above the Earth. The cosmonauts greet their families for one last time from behind a sealed glass panel: children hold up toys and handwritten messages, partners hold up babies, while the cosmonauts sit quietly on the other side of the glass, already wearing their spacesuits.

'I can smell the station,' Lazutkin later recalled of his arrival on board Mir. 'I thought it would smell much worse, but it's virtually normal. But it leaves a taste on your tongue.' The body becomes a basic set of processes in empty space. In 1997 Lazutkin used a small dinner knife to cut through some cables and save a section of the space station after it had been struck by an unmanned supply craft sent from Earth. The further into space we go, the more we learn about ourselves.

NEBULAE

24 September – 23 October

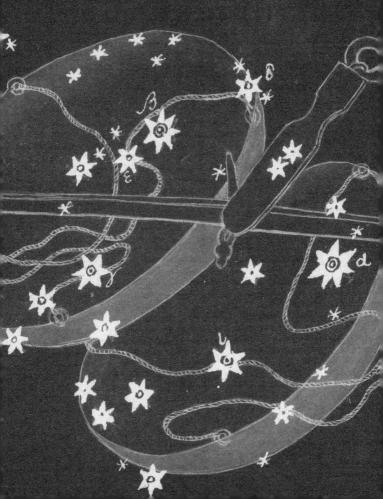

NEBULAE

24 September – 23 October

Also known as:
'Formless' or
Snowflakes

Body Part:
Kidneys and Lumbar region

Areas of Influence:
Distance, Refraction
and Patience

I

On a back road in South Kensington a white panel truck swings by – on the tailgate is a sign: 'NOTHING COMES CLOSE'. Observation is a concentrated study in loneliness – conducted at a distance. The light barely touches the surfaces of our eyes.

Telescopes reveal space as an endless state of unresolved matter: unshaped, unconnected and unrecognised. The effect of any lens is to emphasise the isolation of the eye. Seen through polished glass, stars take on the visible form of snowflakes. Precise yet indistinct, random and symmetrical, individual snowflakes exist between being and non-being, space and materiality, form and void. Throughout the seventeenth century, natural philosophers focus as much upon snowflakes as they do the stars. Johannes Kepler writes a short treatise on the six-cornered snowflake; René Descartes concerns himself with snowflakes in his *Meteorology*; and the Italian astronomer Giovanni Domenico Cassini draws six-pointed water crystals as though they are exotic leaves. Six-pointed stars, as brilliant as snowflakes, begin to appear on charts of the heavens and in astronomical illustrations. The connection seems a natural one: so many stars to discover and document – so many snowflakes left unstudied. Kepler even finds one clinging to his coat in Prague.

Sir Robert Hooke catalogues stars and snowflakes with equal attention. Records indicate that he considers snowflakes to be the imperfect remains of an ideal geometric form, buffeted by storms and damaged in their fall to Earth. Hooke had come to this conclusion by comparing the pattern adopted by each individual snowflake with the more regular form of urine crystals prepared under laboratory conditions. It is the tragedy of metaphysics that it can never shake off its dependency upon material existence. But for their physical battering in the atmosphere, snowflakes would retain the same perfection as frozen urine droplets at the moment of their creation.

A delicate latticework of emptiness and form, a snowflake is something physically located between presence and absence – water as dust. Seen through the eye of a lens, the light from a star is the dust of a star.

II

On 9 September 2009 stargazers report a mysterious streak of light trailing across the night sky. 'I just watched the shuttle and station flyover (8:40 PM CST 9/9/09) and was surprised to see that the shuttle was sporting a massive curved plume,' one observer posts from Wisconsin. NASA later confirms that the space shuttle *Discovery* had dumped ten days' worth of astronaut urine after having undocked from the International Space Station.

Microgravity causes fluid to distribute itself evenly through the body. The kidneys react to this, obliging astronauts to relieve themselves within two hours of takeoff. The space toilet is usually one of the first devices to be activated on shuttle missions. Once jettisoned into space, the urine immediately freezes into a cloud of tiny ice droplets. In the unfiltered sunlight, the ice crystals sublimate directly into vapor and disperse spectacularly into the inky void.

According to Apollo astronaut Rusty Schweickart: 'the most beautiful sight in orbit, or one of the most beautiful sights, is a urine dump at sunset, because as the stuff comes out and as it hits the exit nozzle it instantly flashes into ten million little ice crystals which go out almost in a hemisphere, because, you know, you're exiting into essentially a perfect vacuum, and so the stuff goes in every direction, and all radially out from the spacecraft at relatively high velocity... It's really a spectacular sight. At any rate that's the urine system on Apollo.'

III

It's another fine day in the Solar System. The sunrise is always an optical illusion. What would it mean if it didn't happen tomorrow? In a universe of predictable rotating parts, the Sun is always rising somewhere. Time is local. In a solar eclipse the Sun is precisely in line with both Earth and Moon. Eclipses become a form of public entertainment in the eighteenth century; an intellectual tool of the Enlightenment, their occurrence marks the transition of astronomy into popular culture, which is the mythology of the modern age.

'The Eclipseometer' offers a printed chart for following 'the course of the eclipse of the Sun, every instant of the duration at London, which will happen the 22nd day of April 1715' without damaging the reader's eyes by staring directly at the phenomenon itself. A 'moveable piece' is used to trace the Moon's motion *according to the Computation made by Tables drawn from S.r Isaac Newtons Theory*. The Sun is tinted a lemon yellow. The first to predict how its shadow would fall across the English mainland according to Newton's principles is Edmond Halley. His map depicting the 1715 eclipse is published by John Senex of Fleet Street '*that the sudden darkness, wherein the Starrs will be visible about the Sun, may give no surprize to the People, who would, if unadvertized, be apt to look upon it as Ominous, and to interpret it as portending evill to our Sovereign Lord King George*'.

John Senex had begun as an astronomer and cartographer; he was also an engraver, publisher, surveyor and geographer to Queen Anne. He produced globes and astronomical prints as well as maps of stars, the Sun and Moon between 1712 and 1740. Copperplate engravings were expensive to make so the market for such items grew large over time. Senex's widow sold his original plates, meaning that copies of his work were still being printed long after his death.

Predicting an eclipse reveals the mechanics of the solar system. Everyone can see the Sun, Moon and stars – cheap broadsheets bring them closer than kings, pictured with 'the greatest exactness that can ever be expected'. John Senex publishes English renderings of texts in Latin by Newton and Halley, particularly the latter's 'A Synopsis of the Astronomy of Comets' in which he demonstrates that the parabolic orbits of three recorded comets in 1531, 1607, and 1682 are so similar that they must be periodic returns of the same body. 'Hence,' he writes, 'I dare venture to foretell, that it will return again in the year 1758. And, if it should then return, we should have no reason to doubt but the rest must return too.' While Halley's Comet reappeared as predicted, the broadsheets hawked on the English streets advertised others – such as the one whose *'tail was 200 times hoter than iron'*, expected to return sometime *'in 2225'*.

IV

Snowflakes form themselves into clouds. William Herschel and his sister Caroline raise up within an imposing wooden frame a telescope boasting a 40-foot focal length. Together they catalogue 2500 nebulae, the vast majority of which have never been seen before. These missing pieces of the celestial mechanism appear as milky clouds floating in space: greater magnification resolves them into clusters of stars, requiring new shifts in classification. According to a nineteenth-century account: 'The nebula about 10° west of the principal star in Triangulum is supposed by Sir William Herschel to be 344 times the distance of Sirius from the earth, which would be the immense sum of nearly seventeen thousand billions of miles from our planet.'

The imposing wooden framework supporting the Herschels' telescope starts decaying. In 1839 William Herschel's son John takes an early photograph showing the exposed scaffold – the mixture of salts used in processing the image fixes it against the twilight. John Herschel is later credited with inventing the word 'photograph'.

In 1845 the third Earl of Rosse completes work on the 'Parsonstown Leviathan' at Birr Castle in County Offaly:

a reflecting telescope so large that a contemporary account records how the 'late Dean of Ely walked through the tube with an umbrella up.' Slung between 'two lofty castellated piers 60 feet high', a complex arrangement of chains, pulleys and counterweights allows for the closest possible study of nebulae via a massive 72-inch mirror. Visitors pose in the mouth of the telescope as if emerging from a giant chimney. Reflected light reveals for the first time spiral nebulae so compressed by distance that 'a snowflake would utterly conceal them'. A familiar eye busies itself making familiar shapes out of nebulae. Names suggest resemblances to other common forms: Crab, Fish Mouth, Horse Head. The cosmos will always resolve itself into heterodoxies – how else can it remain whole, or coherent?

V

A camera is a dark room filled with air, surrounded by light. An eyeball is a translucent sphere filled with fluid, surrounded by muscle. Cameras notice things that the eye does not: it happens over time. The more a photographic plate is exposed, the more light is accumulated upon it. On the night of March 26, 1840 John W. Draper takes the first successful photograph of the Moon from a rooftop observatory at New York University: a daguerreotype spattered and blotched with age and chemical changes, the pale crescent on an even paler disc is a result of the plate being exposed for thirty minutes – which turns out to have been too long. Experiments with exposure times and a different arrangement of lenses will produce much clearer images.

Cameras reverse everything, staining paper with light: the image only gets in the way. Nebulae hang like specks of dust in the darkness. 'After the straining mind has exhausted all its resources in attempting to fathom the distance of the smallest telescopic star, or the faintest nebula, it has reached only the visible confines of the sidereal creation.' So begins the section on 'Infinite Space' from *Curiosities of Science Past and Present*, authored by John Timbs and published by Kent and Co. (Late Bogue) of Fleet Street in 1858. It is the shortest entry in the entire book. 'The universe of stars,' Timbs concludes in the second of only two sentences devoted to the subject, 'is but an atom in the universe of space; above it, and beneath it, and around it, there is still infinity.' By the middle of the nineteenth century, astronomy has already taken human sight to the very edge of the eye. In 1880 John W. Draper's son Henry takes the first photograph of the Orion Nebula: a soft clumsily arranged accumulation of light over a fifty-minute exposure time.

In 1882 Charles Scribner and Sons of New York publish the *Astronomical Drawings* of the French illustrator Étienne Trouvelot. Plate XV in the collection shows the 'Great Nebula in Orion' from a study made between the years 1875 and 1876: a graceful double arc of dust and light that Henry Draper's camera did not detect. Trouvelot's painting speaks directly of distance, whereas Draper's photograph awakens a sense of time passing. With the introduction of the camera, astronomy is even less of an art and more of an industrial process. Mounting a camera on a telescope so that it can track a galaxy across the night sky will eventually capture more stars in greater detail than the human eye can ever retain at one time. A ten-second exposure reveals 200 stars – ten hours may show over two thousand. Photographing the sky means that telescopes can see more without necessarily being larger or more powerful. Observatories become hooked up to laboratories, processing and printing out images that can fit onto a page, which means that they can also fit into the eye.

On 6 August 1961 Cosmonaut Gherman Titov takes a camera into orbit on a twenty-five hour mission, photographing the Earth from a porthole in space for the first time. He sees the cosmos formed into a 'black dome' above him and the planet below is transformed into an unfocussed play of light. 'The colors were extraordinary – vivid, yet tender,' he later writes, 'and the light streaming through the cabin carried a strange shade as if it were filtered through stained glass.' Historically cosmonauts' urine was always brought back to Earth for chemical analysis. An external examination of the Mir space station conducted during the mid-1990s identifies a large number of 'flake depressions' that were thought to be caused by frozen slivers of waste water circling the Earth at high velocity and then striking the hull.

PALOMAR

24 October – 22 November

PALOMAR

24 October – 22 November

Also known as:
'Mr Palomar' or
The Temple

Body Part:
Genitals

Areas of Influence:
Transformation,
Instability and The Will

I

Universal Pictures' 1941 production of the *The Wolf Man* starts with Larry Talbot returning to his ancestral home after a long stay in the United States. The film is set on the Universal back lot, which means that it could be taking place anywhere, which is also absolutely nowhere. Lon Chaney Jr. plays Larry, so he has an American accent, and you know something bad will happen to him – like turning into a wolf when the Moon is full and trying to kill the woman he loves. His arrival is immediately followed by that of a large wooden crate sent from London, marked 'GLASS' and carried into the room with great care. 'I believe it's a new part for the telescope,' a servant informs Claude Rains, who plays the master of the house and Larry's father. This is not a good sign.

Astronomy is another form of cinema. Time is fragmented and extended. Matter becomes light in motion. The camera remains fixed, looking outwards into the darkness, while the earth moves beneath our feet. For greater stability, Sir Robert Hooke builds a telescope into the two storeys of his house, the tube emerging from the roof like a chimney, while he observes the star *Gamma Draconis* from the ground floor.

What are the invisible forces that turn a man into a wolf? Larry discovers that the delivery of optical gear is destined for the observatory built by his father into an upper floor of the ancestral home. Its arrangement of star charts, globe and telescope pointing at an open window repeats the painted backdrop Georges Méliès used at the start of 'A Trip to the Moon'. Méliès understood what the camera sees, which also means he understood how one thing can be transformed into another. Meanwhile Larry helps to calibrate the new part for the telescope. 'Where did you learn such precision work?' Larry's father asks, inspecting the results of his labours. 'Optical company in California,' Larry replies simply. 'We did quite a job on that Mount Wilson Observatory.'

In 1941 the Hooker telescope at Mount Wilson, northeast of Los Angeles was, at 100 inches, the world's largest aperture telescope. There was also a 60-

inch version, plus a solar telescope and a 60-foot solar tower, all located at the same spot. The observatory's founder George Ellery Hale suffered from insomnia, depression and other neurological complications that led to stays in a Maine sanitorium and his eventual resignation as director.

Perhaps the unseen power capable of transforming a man into a wolf at night does not originate with the Moon, whose phases may simply provide a way of keeping track of the change. Larry's father asks if he's interested in astronomy. 'Not especially,' Larry shrugs. 'I'm all right with tools. In fact, I've done quite a little work with astronomical instruments, but when it comes to theory, I'm pretty much of an amateur.' You'd think that a wolf man would have at the very least a secret interest in astronomy.

Sir Robert Hooke watches *Gamma Draconis* over a handful of nights through the telescope he had built into his house, draws some erroneous conclusions from his observations then gives the whole enterprise up. 'All astronomers are amateurs,' remarks Larry's father. 'When it comes to the heavens, there is only one professional.' By the time Larry grows claws and fangs and starts ripping the moonlight to shreds, he will be past caring.

II

What makes a man into a wolf? Sometimes frenzied and joyless in his work habits, George Ellery Hale compulsively drives himself to initiate bigger and ever more powerful observatories. Anxiety and depression take the form of a 'demon' he feels rising up within him, accompanied by pains in his head and a ringing in his ears. In between the sleepless nights and days when he finds it impossible to concentrate, he suffers from violent dreams – often waking to find that he is attempting to climb the picture frames on the bedroom wall. Letters to his wife, according to one account, reveal 'a highly disturbed inner man whose thoughts combined repressed sexuality and intense guilt'. His greatest sleep-deprived vision takes shape at the centre of the plateau at Mount Palomar, where the 200-inch Hale telescope still bares itself to the night sky, the observatory dome parting its lips for yet another deep penetration of space.

The Spanish named it 'Palomar' after their word for a dovecote. The local Payómkawichum Indians regarded it as a holy place. Hale begins calling for a giant telescope to be mounted there in 1928. Located in San Diego County, California, Mount Palomar has strong connections with both space exploration and occultism – astronomy has always been, above all else, a structural expression of the unreal: which is to say, the unseen, the undetected and the limitless. The Agapé Lodge of Aleister Crowley's OTO, founded in Pasadena in 1930 for the practice of 'sex magick', considers Palomar Mountain to be the sacral chakra of the Earth. As this is the nerve cluster associated with pleasure and passion, they quickly establish a temple there. Connecting both the observatory and the temple with Pasadena is Agapé Lodge member 'Frater T.O.P.A.N.' – otherwise known as John Whiteside Parsons.

As well as being an adept, Parsons concerns himself with designing and building rockets as part of the Jet Propulsion Labs at the California Institute of Technology. Jack Parsons will later have a crater on the Moon named after him. As well as running the JPL, Caltech is also responsible for the Mount Palomar Observatory with its three giant reflector telescopes and their Art Deco domes. Meanwhile, on the southern slope of Mount Palomar, lives the theosophist and astronomer George Adamski, who runs a small lunch counter called the 'Palomar Gardens Café'. Assuring the public that 'I am not and never have been associated with the staff of the Observatory', Adamski later reveals that he has been using telescopes to watch and photograph flying saucers as they fly over the mountain. He will later claim to have made 'personal contact with a man from another world' in a neighbouring desert location at 'about 12.30 in the noon hour on Thursday, 20 November 1952'. On Tuesday 17 June in the same state and the same year at 5.08, Jack Parsons meets his end in a devastating explosion at his Pasadena home. His right forearm is completely severed by the blast, which breaks his other arm and both his legs, tears a large hole in the right side of his jaw and reduces his shoes to leather scraps.

III

The dream architecture built into the observatories of the early twentieth century was previously replicated in such structures as the *Temple à la Pensée*; conceived as a painting by François Garas around 1900 in a harmonious palette of blue shades, its serene cupola and minaret-topped tower point towards the darkening expanse of human reason. Seeming to grow directly out of a jagged mountain plateau, the temple aligns itself with a distant light gleaming just above a turbulent horizon.

In 1919 Soviet artist Vladimir Tatlin begins work on planning the open geometric arrangement of spiral forms for his *Monument to the Thir▪ International*. Tilted towards the firmament, its slanting tower was intended to extend the Earth's axis out into space along a rigorously modern ideological trajectory. The constituent parts of this gigantic sloping pylon are designed to rotate at different speeds – from once in a day to once in a year. Projectors on overcast days would flash cinematic texts onto the clouds from this complex glass and steel structure. Neither of these buildings is ever realised – François Garas even refuses to practice architecture because he thinks it too practical and boring.

The Hale telescope at Mount Palomar has a huge mirror at its heart, cradled in a nest of girders. The mirror began in 1934 as a perfect 200-inch Pyrex disc, which took seven hours to cast and ten months to cool in a specially designed oven. The grinding and polishing of the disc, interrupted by World War II, took a further eleven years: over five tons of glass had to be removed by thirty tons of abrasive matter before it was considered ready for mirroring. One ounce of vaporised metal was enough to cover its entire 30,000-square-inch surface; concentrated starlight passes through a forty-inch hole in the mirror's centre to reach the observer's eye. Tube and dome can move independently of each other, so that the telescope is able to cover the entire sky. The mirrored disc looks like a giant snowflake when it catches the light.

Observatories grow larger and more permanent, outliving in their development and construction the astronomers who initiated them. George Hale died in 1938, disturbed and confused by a cerebral haemorrhage, ten years before Mount Palomar Observatory was ever completed. Although not due to be operational until 2025 at the earliest, the Giant Magellan Telescope in La Serena, Chile already exists as a logo – you can buy it printed on a duvet cover.

IV

The brilliance of light coming from an object is traditionally depicted in comic strips as a series of straight lines radiating from an unseen centre. Sound is represented as an ever-widening series of concentric circles also emanating from an unseen centre. Wavy parallel lines rise from a pot of boiling water to indicate heat – or even something that smells bad. Imagine wavy parallel lines coming off a light bulb, or concentric circles emanating from an unwashed dog, and you can see how well cartoonists are able to convey the physics of invisible forces.

V

Unseen unless refracted through a prism, the chemical composition of light transforms astronomy into 'astrophysics'. Glowing plasma fans out into a 'bright line' spectrum, more clearly divided than any eighteenth-century colour wheel, allowing the English astronomer Norman Lockyer to announce the discovery of the 'sun element' helium – also found on Earth in radioactive substances. The stars reveal more than their light.

Invisible forces change everything. Following the violent death of Jack Parsons, close associate and JPL cofounder Frank Malina gives up working with rockets and becomes an artist. 'The incandescent bulb and the fluorescent tube are today an inexpensive and reliable source of light,' he writes, 'and silent electric motors of small dimension provide a ready source of rotational motion. In addition, the development of modern plastics has made available transparent and translucent plates which are lighter in weight and much less fragile than glass.' Out of these materials he develops and patents the 'Lumidyne' system: motorised artworks of moving Plexiglas parts projecting abstract visions with titles like *Orbiter*, *Sun Sparks* and *Cosmos*.

In the late 1920s an engineer at Bell Telephone Laboratories named Karl Jansky is asked to find a cause for the unexplained static occurring on shortwave radiotelephone links and minimise it. Unable to identify the source of the hiss, he constructs a 60-foot rotating antenna mounted on four wheels taken from a Model T Ford. He breaks the signals detected from this makeshift device down into three types of sound: the atmospheric discharges of nearby lightning, the clicks of distant thunderstorms and a steady hiss coming from the Milky Way. Jansky's findings are published in 1933 as 'Electrical Disturbances Apparently of Extraterrestrial Origin'. His application to Bell Telephone for funds to build a larger antenna in order to conduct a more detailed study is turned down as not being cost effective.

As radio is a weapon of communication and control, research into its applications as an astronomical instrument was, like the grinding and polishing of the Hale telescope's mirror, suspended for the duration of World War II. The Lovell radio telescope went online at the Jodrell Bank Observatory outside Manchester, England in 1957. The mechanism used to steer this giant parabolic dish uses a system of motorised parts taken from the gun turrets of British warships HMS *Revenge* and HMS *Royal Sovereign* to keep it steady even under the stress of storms, high winds and heavy rain.

When mirrors and lenses no longer extend the human eye, the ear takes over – and unseen stars are heard first. Optical and radio images are superimposed to create dense clusters of information that pass through a widening bandwidth. The prime mover behind all celestial activity remains only partially glimpsed, however. *'Millions beyond the former millions rife,'* exulted the Revered James Hervey, pointing towards the eighteenth-century stars. *'Look further: --- Millions more blaze from remoter skies.'* In the early years of the twenty-first century Microsoft's 'Starfield' screensaver plunged users into an endless journey through the cosmos, past receding lines of stars towards an algorithm-generated vanishing point that will never arrive.

LANDSAT

23 November – 21 December

LANDSAT

23 November – 21 December

Also known as:
'Recon' or
The Surveillance Satellite

Body Part:
Hips and Thighs

Areas of Influence:
Circumscription, Movement
and Education

I

As the darkness takes shape around it, the Earth assumes a bright formless radiance. One of the first to catch a secret glimpse of the planet's blue halo was cofounder of the Polaroid Corporation Edwin Land. He developed the optics used in the Lockheed U2 spy plane, including a 12,000-foot spool of specialised high-resolution film and the massive 500-pound Hycon 73B camera: a bulky grey custom-built device that guaranteed sensitivity over distance on an industrial scale. The U2's black unmarked fuselage effectively gave birth to the first reconnaissance satellite – the wings merely provided lift-off. Officially the plane did not exist, and the pilot carried no identification papers. Land's airborne imaging system was designed to skim across the outer edges of our atmosphere inside an anonymous motorised glider following vast 4,000-mile arcs that ignored every international zone and national border that lay below it.

Months after America's U2 surveillance programme began, the Soviet Union launched Sputnik 1 from its secret launch site at an isolated railhead in the middle of a barren desert. The place did not appear on any map, and its mail address was a Moscow post office box number. Later it took the name Baikonur, referring to its supposed proximity to a non-existent lake, to further obscure its location. High-altitude planetary reconnaissance consequently began with a satellite that didn't exist being photographed by a plane that didn't exist over a location that didn't exist. 'Soviet Territory has become the Shores of the Universe,' Sputnik's designer Sergei Korolev would exult under a false name in a New Year's edition of *Pravda*.

How else to see this blue halo – since no one else could – except as a sign of the Earth's inner life? Sputnik's presence was more auditory than visible, with people across the planet tuned in to its constant beeping. The Shores of the Universe existed in the ear rather than the eye. Sputnik was a disaster and a media event at the same time, making space audible but keeping the contours of the Earth a closely guarded secret. The world had changed, and people were only just noticing it – the sign of a culture already fascinated by its mass communication systems and how they operate.

To confront the unconscious as an environmental fact was to encounter a hollow world. This was given material form at the centre of the 1964 New York World's Fair, where the stainless steel Unisphere offered a shimmering representation of an empty world as seen from space, its oceans and continents splayed out as flattened shapes rolled out across its non-existent surface. Shining steel rods trace unbroken orbits around these formations. During a radio interview given with the BBC around the same time as the World's Fair, Marshall McLuhan referred to the Earth as 'an old camp thing' encircled by communication satellites: isolated in the emptiness of space as if it were some outmoded machine.

'Camp' was and is the ecstatic celebration of the absent and unattainable: Hollywood captured in a naked key light and a handful of sequins: the grand drama of an amateur stage. The planet McLuhan described as 'camp' was in the process of disappearing even as it was being observed. Captured in the lenses of high-altitude reconnaissance cameras, a new Earth had emerged: one of folds and creases, rivers and plains, geometric assemblages of buildings and invisible frontiers. Its terrain was scanned in continuous sweeps, processed and laid out in photographic prints to be studied by unseen eyes.

II

A transept in technological time:

1912 – 1926
There are only title cards that do not move but suggest that the Earth is encircled by some kind of ring or early form of orbit. Later things start moving – a darkened planet appears against a mottled firmament; the ghostly outline of a smiling man's face flickers and flares. 'Carl Laemmle' appears in white Italianate script at the bottom of the screen. Later still a gigantic biplane circles the globe in silence, leaving behind it a wispy trail that spells out 'UNIVERSAL PICTURES' as the vapour separates under the pull of gravity – the 'U' at the beginning and the 's' at the end distort and bend themselves around the world.

1926 – 1936
The Earth is a smooth ball turning slowly against the clouded heavens. Shadows loom. The continents are all in relief. A monstrous single-engine plane follows the equator, remaining fixed above the Pacific Ocean as it disappears into the East. 'A UNIVERSAL PICTURE' appears in a diagonal wipe across Africa.

1936 – 1946
A chrome sphere radiates transparent shafts of light into a black firmament. Seeming to emanate from some unseen centre in space, their rays twinkle and

coruscate. Five-pointed stars rotate in the fixed cosmos – everything looks as though it were made of glass. 'A UNIVERSAL PICTURE' appears as a three-dimensional band orbiting the chrome sphere, the word's reflection traversing its equator.

1946 – 1963

The Earth is turning against the starry cosmos. North and Central America stand out against the featureless grey ocean. 'Universal International' appears in flat stylised lettering that casts an equally flat shadow on an unknown surface. Clusters of stars give out a soft, slightly blurry light. The scene fades to deep black. A Technicolor variant shows the same landmasses as textured brown forms on a featureless blue ocean that now glows with reflected light from an undisclosed source. The stars disappear, and there is only the deep blackness. 'Universal International' is in the same stylised lettering but no longer throws any kind of shadow. The scene fades as before.

1963 – 1990

The stars rush forwards in dreamy profusion. The Earth suddenly emerges from the inky black distance, turning slowly as it comes into view. Pale spirals of glowing blue dust entwine themselves before it. 'A UNIVERSAL PICTURE' appears in textured gold letters with white capitals. The stars move in repeated patterns through the darkness and blue dust. The Earth spins serenely as the scene fades.

1991 – 1997

A flash of light is caught between the curve of a darkened blue world and the ragged violet nebulae of space. The light flares as the perspective changes; and the light disappears to reveal a bright chaos of stars and glowing dust clouds. Landmasses stand out against a blue ocean as the light comes up. Gigantic yellow and black letters suddenly reveal themselves: orbiting counter to the Earth's rotation, they flatten out to form the word 'UNIVERSAL', which hangs in space while the world keeps turning. We reach the Americas, and the scene fades.

1997 – 2012

A rim of golden light forms in the darkness: the planetary horizon shapes itself – the glow intensifies. Powerful shafts of light suddenly burst out from the darkened sphere, breaking up the horizon. Is this the final destruction of an unstable planet torn apart by some inner power surge? The intense shards of light take on the shape of the recognisable landmasses. Out of nowhere 'UNIVERSAL' slides in from the right-hand side of the screen – huge white letters that throw golden shadows in space, which flatten out as the Earth recedes into the background. Things are calmer now: the oceans are a rich deep blue, the Americas are a gentle symphony of greens and yellows and browns. Even so the Earth emits a white spiky nimbus of radiation as if generating its own light.

2012
A suborbital view of the Earth glowing against black space: running backwards at rapid speed across blue ocean and brown landmasses streaked with white clouds – everything is happening too fast to get a fix on an exact location. We now seem to be accelerating away from the Earth, concentrated sunlight flares from a corner of space – and here comes 'UNIVERSAL' again in letters of silver and gold so huge that the word can wrap itself around the entire planet. It has the intensity of a sunrise. There seem to be fires burning throughout the world, as the word sails serenely by. The fade to black is so quick we barely notice it.

This is the way the world turns.

III

Picture this scene from junior science class: a magnetised iron needle floating on a strip of white paper or a sliver of cork in a shallow dish of water. Left to its own devices, the needle will align itself with the north and south poles of the planet – assuming no anomalous magnetic interference in the form of ghosts, evil spirits, meteorite fragments or other magnets. The floating needle is timeless and simple and something you can rely on.

The magnetised needle is a tiny star by which to navigate, no matter where on Earth you might find yourself. A compass was meant to travel the globe; the true meaning of its constancy is only revealed through its mobility.

All terrestrial globes begin and end at the equator, which is little more than a moment: a line to be crossed – the continuous point at which the sphere is widest. Trade and exploration connected the globe up with the magnetic compass needle. The equator became a line of conquest on a map: a virtual location where navigation has turned into ideological conflict.

IV

A printed globe of the night sky made in the sixteenth century by the great Johannes Schöner shows the Zodiac spread out against the darkness of Heaven. For observation purposes, the Earth's horizon is located across its centre: an equator in the sky. Schöner's celestial globe can be glimpsed in the background to Holbein's *The Ambassadors*, alongside other objects that represent the *quadrivium* of liberal arts: Astronomy, Mathematics, Geometry and Music. Leaning against the ranked shelves of knowledge, formal and perfectly at ease, stand two Renaissance astronauts on a mission to explore diplomatic space. Protected by the rich and heavy robes of their office, hands resting either against a hip or the top of a thigh, are Jean de Dinteville, the French ambassador to England in 1533 and his friend, Georges de Selve, ambassador to the Venetian Republic and the Holy See. Before them in the immediate foreground is an anamorphic projection of a human skull, stretching across an unseen plane as if it were viewed from space. When seen at the correct angle, the distortion disappears; and the cranial orb is clearly revealed.

The catalogue for an official auction of Elvis Presley's belongings held at the MGM Grand Hotel & Casino, Las Vegas in 1999 contains the following entry: 'Item C228 ELVIS PRESLEY'S GLOBAL BAR: Unique gentleman's bar in the shape of a globe with ornate inlaid features. Four page yellow receipt and contract dated May 27 1975. 13¼ x 8½ inches receipt. 29 x 28 inches globe.' Elvis and every astronaut since Neil Armstrong has walked onstage to the opening bars of Richard Strauss's *Also Sprach Zarathustra* – but only Elvis shook his hips and danced.

V

Since the launch of Sputnik 1, over 6,000 functioning satellites have been sent into space. Today Google Earth, with its attendant processing systems, has installed itself as a suborbital selfie stick – perpetually focussing in and out of the planet's surface, hopping from detail to detail. 'We're looking at ourselves,' remarks an astronomer from the European Space Agency, 'and that's the biggest thrill.' At the White Sands Missile Range in New Mexico, the Defence Advance Research Projects Agency, otherwise known as DARPA, has developed a 90-ton surveillance telescope capable of tracking 10,000 objects in motion around the globe at a time. Its function, according to one report, is to keep track of 'space junk, missile tests, dead rocket parts, foreign satellites of unknown intent and uncharted asteroids coming in for the kill.'

In 2008 a group of students in Catalonia released a cheap latex weather balloon filled with helium carrying sensors and an everyday Nikon digital camera. It rose twenty miles above sea level, where the homemade satellite photographed the blackness of space, the glowing edge of the Earth with its unbroken layer of cloud cover. *Popular Mechanics* magazine later published advice on 'How to Launch a Camera into Space' using a weather balloon, a parachute, some aluminium foil, hand warmers and a stack of old newspapers. In choosing a suitable location for take-off, the article suggested, 'be sure not to launch near military installations. Stay at least 100 miles away.'

TSANDER

22 December – 20 January

TSANDER

22 December – 20 January

Also known as:
Cosmopolity or
'The Ascension'

Body Part:
Knees and Bones

Areas of Influence:
Disappearance,
Suspension
and Aspiration

I

Ascension is a practical art: the spiritual resolution of altitude and elevation. The impulses and motivations behind it are easy to understand. 'I was always awed by flight,' Edgar Mitchell of Apollo XIV revealed in an interview. 'When I was just a young lad a barnstormer flying a World War One airplane landed on our farm, and Dad helped him refuel, and I got a ride and he took me for a circuit of the field, and that was my first airplane ride at about four years of age.' Mitchell believed in the existence of UFOS and was the sixth person to walk upon the Moon.

In 1811 Madame Sophie Blanchard, Europe's first professional female aeronaut, cuts her balloon loose for an aerial display over Milan. A strong wind blows her across the Apennines, almost as far as the coast of Genoa. Eventually Madame Blanchard lands just outside a mountain village where the inhabitants kneel before her, believing she is an apparition of the Virgin Mary come down from Heaven.

In October 1899, before the Wright Flyer had ever lifted itself from the ground, a teenage boy uses a homemade ladder to climb into the branches of a cherry tree in Worcester, Massachusetts. Looking out across the fields the seventeen-year-old Robert Goddard imagines 'how wonderful it would be to make some device which had even the possibility of ascending to Mars, and how it would look on a small scale, if sent up from the meadow at my feet.'

The Emperor Napoleon appoints Madame Blanchard as his 'Aeronaut of the Official Festivals'; later, after Napoleon's demise, King Louis XVIII names her 'Official Aeronaut of the Restoration'. During a public display at the Tivoli Gardens in Paris in 1819 a stray spark from a firework sets her balloon on fire, sending it crashing into the roof of a house on the nearby rue de Provence, where Madame Blanchard falls to her death.

Robert Goddard's theoretical writings on liquid-fuelled rocket flight make him a cultural hero in the

Soviet Union during the early years of the Revolution. A public debate organised by the Society for the Study of Interplanetary Travel at the Physical Institute of Moscow State University on 'The Truth about the Dispatching of Professor Goddard's Projectile to the Moon on 4 August 1924' attracts such large crowds that mounted riot police have to be called in to keep order.

II

I was barely a teenager when I visited the Soviet Union. It was a great summer: the Sun shone on everyone. I smoked my first cigarette: a Soviet brand called OPAL I think. Kissed a girl for the first time. Was taken to my first gay bar in Leningrad – one of the few places still open behind closed doors that late at night. I still remember the black-market traders who would pass quietly down the tourist hotel corridors, insistently calling you 'friend', and the beautiful CCCP party pins in brightly coloured enamel. Most of all, on its green patch of mowed grass against an open blue sky, I remember the Monument to the Conquerors of Space near Prospekt Mira in northeast Moscow. Riding a sculpted 350-foot plume of smoke and condensation, a rocket rises above the city traffic, commemorating Yuri Gagarin's 1961 orbit of the Earth. That seems such a long time ago now. Around the monument's base a cosmonaut ascends the Hegelian steps, supported by a heroic frieze of men and women engaged in increasingly refined forms of labour: agricultural, industrial, technical and scientific.

The pioneers of space hold up models of Sputnik 1, Vostok, spools of punched computer tape. Looking back, it has the air of an ancient mausoleum: one that has long since outlasted the hands that designed and built it. Titanium plate and worked metal gleam in the sunlight, while Laika, the first dog into space, gazes modestly down at the ground.

III

In 1927, to mark the tenth anniversary of the October Revolution and the seventieth birthday of rocketry pioneer Konstantin Tsiolkovsky, the 'First Universal Exhibition of Models of Interplanetary Apparatus, Mechanisms, Instruments and Historical Materials' opens in Moscow. Coordinated by the Soviet Association of Inventors, an informal organisation of students and workers dedicated to the cause of Soviet cosmism, it collects together for the first time plans, prototypes and artwork to aid humanity in its ascension to the outer universe. 'By taking a pair of steps,' one of its organisers remarks, 'I crossed over the threshold from one epoch to another.' In which direction can these steps have been taken except upwards and into space?

One of the items on show is a model of a 'space plane' designed by Fridrikh Tsander – this would use its multiple wings and engines to reach the outer edges of Earth's atmosphere, at which point it would convert into a spaceship powered by aluminium and magnesium parts taken from the fuselage. Its multiple wings, fuel tanks and propeller would be dismantled and thrown into special chambers where they would be shredded, crushed and melted down before being pumped into the engines. A sign indicated that Tsander had designed this 'self-consuming' rocket plane in 1922. Fridrikh Tsander affirmed 'with almost total certainty that intelligent life exists on Mars' and considered the planet to represent humanity's future: the ultimate object of space exploration. He ended all of his lectures and speeches with the same words: 'Onward to Mars!'

Over 10,000 Soviet citizens attend the show. Visitors to 68 Tverskaia Street in Central Moscow, where the exhibition takes place, are greeted by a large diorama of a rocket landing on the Moon. A giant Earth is shown for the first time as a distant prototype: a plaster model outlined against a painted backcloth, rising high above the cloudy white cosmos. Above it was a sign written in a hitherto unseen alien language: 'Cosmopolity invents roads to new worlds.'

IV

Fridrikh Tsander speaks to us from beyond the grave:

In 1920 I took advantage of Comrade Lenin's time and cordiality to tell him in great detail of my plans to build a rocket-powered spaceship that would eventually take mankind to Mars. 'Will you be the first to fly in it?' he asked. The future

129

*of all humanity lies in outer space – it is our spiritual destiny,
handed down to us by Konstantin Tsiolkovsky, who showed
us that rocketry conformed to the Will of the Universe, and
Nikolai Fedorov, who decreed that the common purpose of all
science is to attain the immortality of all people over all time.
Cosmonautics must remain first and foremost a spiritual
discipline. Those who would make up the Cosmopolity of the
future are effectively renouncing their life on Earth. They
must follow a strict vegetarian diet, learn to survive without
sleep, strive to promote social equality and develop a universal
language. Who knows what they will have to undergo, how
they will live and perish in space, in order to open a new
page in the discovery and conquest of the universe? A New
Man must be created in order to conquer this New Cosmos.
Through poverty, sickness and ill health I stuck with this
principle. Fedorov had blood transfusions to prolong his life.
I named my children Astra and Mercury even though they
held me back from my greatest work. I founded the Group for
the Study of Reactive Motion and inspired the work of Sergei
Korolev, the Great Designer, who understood that Sputnik
had to be a single shining silver sphere – he knew that it had
to reflect the light of the stars and that it would one day be
exhibited in the museums of the world. In 1958 while Korolev
lay in a Kislovodsk sanatorium recovering from the physical
strain of launching Sputnik 2, he sent out members of his staff
to find my grave. He later saw to it that a monument was
built to the man who first opened his eyes to the possibility of
rocketry. Onward to Mars.*

V

Ascension is the art of becoming a dot – something barely glimpsed from a distance. Announced immediately after the fact, the Soviet Union's early space triumphs seemed to have taken place in retrospect, leading to the suspicion that they must have been preceded by some terrible failures. Two Italian brothers, Achille and Giovanni Battista Judica-Cordiglia, installed a shortwave radio station in an old German World War II bunker outside Turin to track Soviet and US satellites. Later they claimed to have intercepted the final transmissions from at least four space missions prior to Gagarin's – capsules that veered off course or kept on going into the depths of interstellar space. Later still the brothers announced that they had recorded the sounds of cosmonauts burning up on re-entry. The disputed authenticity of these recordings is not the issue – trainee cosmonauts had already been airbrushed from official photos, and one was removed from all historical records when his flight simulator burst into flames during a training session. The 'Lost Cosmonauts' survive as myth: phantom media presences, linking outer space with radio signals and tapes and finally with death itself – with something that is becoming us but remains other than us.

On 28 January 1986, 73 seconds into its mission, tracking cameras identify a plume of hot gas on space shuttle *Challenger*'s right-hand solid rocket booster, causing a flame to ignite the liquefied oxygen and hydrogen in the main fuel tank. Burning debris rains down on the Atlantic Ocean for an hour after the explosion, which occurs just after 11:39 am. 'It's hard to dazzle us,' President Reagan tells the nation on television. 'We've grown used to the idea of space and perhaps we forget that we've only just begun.' Only 17% of the American population watches the live coverage of the *Challenger* disaster. The major broadcast stations had already cut away, only immediately to cut back again to a taped relay of the event. Within an hour almost 90% of all Americans are watching – but what most people will later recall as a live broadcast is actually a delayed version of events.

An intense dispersal of energy with nowhere to go except everywhere, the plume marks a moment between a forceful presence and an eventual disappearance. It is an undetected silent broadcast – an alien transmission. The violent plume that resulted in the disintegration of the *Challenger* space shuttle is echoed later that same year in the flailing tentacles of radioactive ions emanating from the Chernobyl reactor, pulled around by local weather systems across mainland Europe.

Cosmopolity as a technological projection out into space ends with the diffusion of material being. The immortality of existing in space, the promise at the heart of the Cosmist dream, is experienced as a form of loss. Those who spend extended periods of time aboard the International Space Station produce high concentrations of calcium in their urine. Linking this with the measurable loss of bone density caused by prolonged exposure to conditions of microgravity, one observer commented that astronauts are slowly 'pissing out their skeletons one atom at a time'.

The first cosmonaut changed the world simply by leaving it. On the way to the launch pad at the Baikonur Cosmodrome in Kazakhstan he requested that the bus stop so that he could relieve himself against the right rear tyre. Since then the bus always stops for each astronaut to do the same: the males unpacking themselves from the many layers of their spacesuits and then being sealed back into them again, the females splashing vials of their urine on the sacred black rubber. A Soviet-era film called *The White Sun of the Desert* is screened at the Baikonur Cosmonaut Hotel on the evening before each launch. Every astronaut watches it – they have to. There is no precise reason for this tradition, its origins have become unclear over time. Astronauts also plant a tree at Baikonur and sign their hotel door – all in remembrance of the first cosmonaut. Everyone that follows him now leaves to the sound of 'Grass by the Home', a song about space travellers longing for life back on Earth performed by the state-sanctioned rock band Zemlyane – their name means 'The Earthlings'.

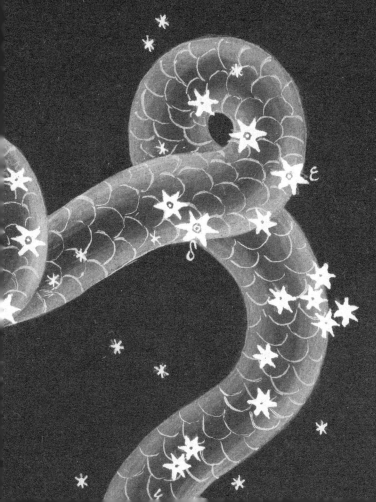

GHIDRAH

21 January – 19 February

GHIDRAH

21 January – 19 February

Also known as:
The Three Headed-Monster,
Monster Zero or King Ghidorah

Body Parts:
Calves, Shins and Ankles

Areas of Influence:
Reflexes, Extremities
and Illusions

I

Life in the universe did not start with reproductive biology but is simply maintained by it. Chemical reactions take place so slowly in the extreme cold of deep space, molecules gently colliding together, free from the intimate pull of gravity. How does any world escape its mineral existence for the type of cellular division that sustains plant and animal life? Sigmund Freud suggested that 'in addition to the drive to preserve the living substance and bring it together in ever larger units, there must be another, opposed to it, which sought to break down these units and restore them to their inorganic state.' The urge to explore space is, in the fullest sense of the term, a death drive – the dominant question might therefore be: how do we leave our human physiology behind?

Getting into space remains a comparatively simple matter compared with the logistics of staying there: we are constantly dodging disaster. Leaving gravity behind still means taking your body with you. December 1954: in a thick cloud of white smoke United States Air Force flight surgeon John Paul Stapp clocks 672 mph in a rocket sled running along a straight track stretching into a dry alkali desert. Snapping to a dead halt in less than a second, Stapp subjects his body to a 43-G deceleration. 'His eyeballs just about popped,' a witness recalls. Stapp is blind for ten minutes afterwards: doctors probing his injuries against a green hospital wall – dust particles have studded and scoured patches of his skin through his flight suit. Stapp is gently lowered back onto a white pillow, sockets turned to a dark jelly.

Anticipating the physical demands of weightless existence, medical experts in the 1950s feared that astronauts would choke to death on food they could no longer swallow. Lack of air movement might cause them to suffocate on the carbon dioxide they were unable to expel. Future cosmonauts were sent for long mountain walks to simulate high altitude and experimented with eating upside down to determine whether food would still travel from the mouth to the stomach. Long-haul space travellers spend months lying on tilted beds with their ankles raised above their heads, their hearts shrinking from blood loss and their temples pounding.

II

During the 1960s outer space was filled with men in rubber suits. The giant monster Ghidrah invades Mars – despite the Martians' advanced technology, he wipes them out in a matter of weeks. A female prophet from Mars warns the Japanese people that the Earth is next. Doom and destruction are near. A meteorite streaks through a clear blue sky. It splits in two, and Ghidrah emerges in an animated ball of yellow fire – a golden, winged dragon with three heads firing powerful bolts of electricity from each mouth. Only Mothra, Rodan and Godzilla working together can defeat the three-headed monster once it destroys the world's cities and tall buildings. Later called King Ghidorah, the creature returns in *Monster Zero*, also known as *Invasion of Astro-Monster*, and *Destroy All Monsters*. Filmed at street level, working the lower legs of the creature's costumes, young male actors as fit as any astronaut stamp their way across the studio soundstage.

Serapis, the Egyptian god of the Underworld who wears a Greek face, is accompanied by a three-headed monster that rules over time past, time present and time future. The Underworld marks the deepest impression of outer space on Earth. A three-headed serpent, representing the development of human philosophy, appears on coins minted in Venice by Giovanni Zacchi in 1536. The Renaissance *tricipitium*, as depicted by Titian, bears a Latin inscription that translates as 'instructed by the past, the present acts prudently lest the future spoil its action'. Ghidrah will always return.

III

Summer of 2014: the Russian space agency loses contact with their Photon-M satellite containing one male and four female geckos destined to be part of an experiment in zero-gravity sexual reproduction. By autumn the lizards are dead, their frozen mummified remains returned to Earth in the probe's landing module. 'It's still too early to talk about the geckos' cause of death,' the agency announces to the press. It was thought that the creatures' relatively low metabolisms and sticky feet would counteract the effects of weightlessness. An experiment involving fruit flies on the same mission is more successful, the insects thriving and breeding.

Published in 1970 by the Barnaby Press as an illustrated paperback, *Outerspace Sex Orgy* claims to be 'the first sex science fiction by Arthur Faber' – which can mean almost anything. The cover photograph reveals outer space to be decorated with stick-on stars, a glowing distant moon and scraps of Reynolds Wrap. Standing against the dark cosmos, three figures study each other with gentle curiosity. Two earthlings – one male and one female – are naked except for curved space helmets trailing air hoses. A nude alien poses in green body paint, a length of Christmas tinsel around her waist – she has her back turned to the camera and pointed red ears. 'FIRST AMERICAN PRINTING, ADULT READING, NOT TO BE SOLD TO MINORS' is printed in bold white letters across her calves and the backs of her ankles.

The first earthbound creature to mate in space was a common house cockroach named Nadezhda, which means 'hope' in Russian. A video camera filmed the entire process. Nadezhda gave birth to thirty-three cockroaches, the first to be born in conditions of microgravity. That was in September 2007. The babies' shells darkened much earlier than usual; and, according to researchers, they grew to 'run faster than ordinary cockroaches, and are much more energetic and resilient'.

IV

I dreamed I saw the stars in my Maidenform bra...

Given the law of large numbers and the amount of bodies that have been into outer space for any extended period of time, the chances of at least one zero-G jack having already taken place are pretty high – and someone knows the details. Spacesuits and close proximity cannot always get in the way. Sex in a spacesuit is always safe; all-over protection, intimately shaped for each individual body, covers everything. Sex in a spacesuit bonds you with yourself. American astronaut Alan Shepard has his skin tattooed with the locations of body sensors to be affixed before lift off, connected through his spacesuit to the capsule's on-board monitoring system that would relay the data back to the physicians at Mission Control.

Ascension imprints itself upon mammalian physiology as hypothermia, dizziness, organ trauma, dehydration, oxygen scarcity and insufficient pressure to support bodily tissue. Atmosphere and gravity form an embrace. Dogs and monkeys would squirm, laced into webbing designed to hold them immobile in their capsules during the early missions. On the 'Man in Space' episode of *Disneyland* Wernher von Braun shows water boiling at blood temperature in a pressurised chamber to demonstrate the effects of high altitude on the human body.

I dreamed I was floating through space in my Maidenform bra...

Before taking off for the outer edges of the sky, high altitude test pilots would breathe pure oxygen for an hour to lower nitrogen levels in their blood. The first pressure suits resembled female corsets pulled tight to prevent the G-force from draining blood from the head and pooling it in the legs. Spacesuits and items of women's lingerie being among the most complicatedly elasticised garments of their age, the distinctions between the two were quickly erased. Wally Funk, trainee in the 'Mercury 13' female astronaut programme, wore her mother's girdle under her flight suit, surviving five times the usual force of gravity in a centrifuge. Playtex Inc. and the Spencer Corset Company won contracts for aviation suits, while the David Clark Company applied their expertise in aerospace protective equipment to bra manufacture.

Learning to move encased in nylon and rubber is 'like trying to adapt to life within a pneumatic tire', according to one account. Astronauts are always in their underwear: their bodies closely bound to space, skin marked by the pressure of webbing and straps. Zips, laces, clasps and seams are delicately hand-sewn into place by seamstresses – the aluminised outer layer shining 'like liquid metal poured over each astronaut's body'.

I dreamed I felt the Earth move in my Maidenform bra...

To preserve the body forever in space is to cancel it out as a physiological process, removing it from time. Functions and definitions are blurred by extreme distance. Made from Lycra and neoprene, the panty girdle worn under Mercury astronaut Gus Grissom's spacesuit to hold some additional drainage tubes in place was the most technologically advanced control garment available at the start of the Space Race – it was bought off the rack at a women's boutique in Cocoa Beach, Florida.

V

All simulation is a form of science fiction: a projection into the future. Aliens encountering NASA's *Voyager 1* space probe, which left our Solar System long ago bearing greetings from planet Earth, will never get to see us in our spacesuits. A male and a female will greet them instead, standing side-by-side, naked and hairless, the man with his hand raised – porn stars representing the entire human race. The message implied in this welcoming scene is that, aside from intergalactic communication, the only other science we seem to have mastered is the full Brazilian waxing. Meanwhile *Voyager 1* is now over ten billion miles from the Sun, having travelled further than any other manmade object in the known universe; it is not scheduled to approach any specific star or galaxy for at least another 40,000 years. Enough time has elapsed since the probe's launch in 1977 for us to assume that the only alien perceptions it has so far confronted have been our own. Tastes change, after all, separating us out across time rather than space.

Over the intervening years we have sent out even more probes, satellites and planetary rovers, transforming space into a sensory download, an extension of our own collective nervous system. This rich flow of information forms itself into weather systems, rock formations, geological drifts and flows. Planetary space is now online: we are connected directly to it. If humans have long since become the sex organs of machines, then simulation is the technological equivalent of pornography. Body parts, sensors and transmitters form communicating spines along which raw data flows like fluid – from limbs to devices and back again.

The simulation of direct experience erases distance – which is something that can never be simulated.

Probes and organs fit together to form a body without a functioning interior: one that inhabits a digital hallucination. Everything seems familiar here, and yet meaning remains elusive. 'We are embracing the Rover,' the Mars Geologist says. 'Technical reports in which we say "the Rover has explored" and "the Rover tested" are rejected because the language is not "scientific", but the Mars Rover acts for us in an emotional sense.' The Mars Rover communicates through an office space like any other – bright desks loaded with workstations, strip lighting glowing overhead. 'It has a personality,' the Mars Geologist continues. 'It's not a machine for us. We attribute moods to it – a personality. "Rover isn't talking to us." It's impossible not to have a relationship with the Mars Rover.' When he refers to 'we' or 'us', he is talking about the other 400 people involved in the mission. At the time of writing, Mars is the only planet in our Solar System exclusively inhabited by robots – but only two of them are currently working. The Mars Geologist describes 'the excitement of coming over the brow of a hill on Mars and seeing a new terrain – a Martian vista that nobody has seen before.' The Mars Rover is the landscape artist of the twenty-first century. Images sent back from stereoscopic cameras are stitched together back on Earth: the result is a composite portrait showing the Martian topography on different days under different light conditions. The black parts of each scene are where there is no component data. 'One of the imaging technicians didn't want Rover to roll over a specific feature because he didn't want a tyre track on it,' the Mars Geologist explains. 'He was keen to keep the site pristine. "Don't drive over that outcrop," he'd say. "Because landscapes should go on forever."'

PLUTONIA

20 February – 20 March

PLUTONIA

20 February – 20 March

Also known as:
The Dwarf, The Nubian
or 'Planet X'

Body Parts:
Feet

Areas of Influence:
Limitations, Fantasy,
Effects

I

For a long time now the astronomer's job has been to show us in the greatest possible detail all of the worlds we cannot possibly inhabit. Is this what we have prepared ourselves for? We live in fantasies of the impossible.

A young medium, known to readers of the book *From India to the Planet Mars* simply as 'Hélène Smith', conducts a series of séances in Geneva between 1894 and 1901. Closely observed by the psychiatrist Théodore Flournoy, Smith will habitually enter a deep trance and 'thereupon mimic the voyage to Mars in three phases'. First Smith will rock the upper part of her body backwards and forwards to indicate that she is 'passing through the terrestrial atmosphere'. Next she holds herself rigidly immobile while crossing 'interplanetary space'. Her descent through the Martian atmosphere is communicated by 'oscillations of the shoulders and bust'. Once arrived upon Mars, Hélène 'performs a complicated pantomime expressing the manners of Martian politeness: uncouth gestures with the hands and fingers, slapping of the hands, taps of the fingers upon the nose, the lips, the chin, etc., twisted courtesies, glidings, and rotation on the floor, etc.' Such motions, Flournoy notes, represent the various ways in which Martians are accustomed to greeting each other.

A drawing by Hilary Knight for Kay Thompson's 'book for precocious grown ups' shows Eloise standing naked in the bawth, talking urgently into a shower attachment. She holds a Dixie Cup to her right ear. A winged spaceship with a transmitting antenna forms itself invisibly around her waist, while unseen children struggle and drown in the bathwater. 'Paper cups are very good for talking to Mars' reads the caption beneath this apocalyptic scene.

On 17 September 2014, during the 751st Martian day, NASA's Curiosity rover photographs the 'Pahrump Hills', an outcrop within the Gale Crater site located near the equator. Nobody wants to see Mars turned into a graveyard, so locations are named after places that already exist on Earth rather than living individuals. Pahrump, for example, is a small town in the Nevada desert where legalised brothels and gambling flourish. It is also where the United States government directs the Martians to make their first landing in the 1996 movie *Mars Attacks!* Shortly after their saucer touches down and its ramp uncurls like a large metal tongue onto the desert floor, the Martians start massacring everyone in sight. Pahrump is reputedly derived from an indigenous name for the region meaning 'Water Rock'. This clearly isn't over yet.

II

A boy reclines stiffly in a plastic chair, which in turn is lying upon its back in the middle of a classroom floor, knees pulled up to his chin, body braced for take-off and the resultant G-force. He is wearing a plastic space helmet – a chalk drawing on the blackboard behind him shows the Moon and the Earth with a rocket's trajectory from the equator. His teacher and classmates look on.

A young girl, obsessed with the American TV show *Lost in Space* since the age of five, plays in a darkened sitting room. The door suddenly opens. 'Mother!' the girl yells. 'Close the door! I'm in another galaxy.' The door closes again.

Sun Ra and his Myth Science Arkestra release an album of original compositions and arrangements called *The Nubians of Plutonia* on the Saturn Label in 1966. The recordings for this long player were made during rehearsals in Chicago sometime between 1958 and 1959. Tracks listed are as follows: (Side One) 'Plutonian Nights', 'The Golden Lady', also known as 'The Lady with Gold Stockings', 'Star Time' (Side Two) 'Nubia', 'Africa', 'Watusa', 'Aethiopia'.

Sun Ra instructs his musicians that they should be very careful what notes they play when performing as their music might carry across intergalactic space, and any discords could adversely influence people and events in other places and other times.

In 1972 the Impulse! Records label began reissuing many titles from Sun Ra's back catalogue, including *The Nubians of Plutonia*, in gatefold sleeves with new artwork. When ABC, the label's parent company, suddenly dropped the project, you could pick up copies of these albums in the bargain record bins for next to nothing. Sun Ra never saw a penny in royalties from these sales, but there must have been thousands of purchases made by people who had never heard of him or his music before but who were attracted by the low price. The continuing impact of these unrecorded sales upon musical development over the next century would run incalculably deep.

III

The astronomer Clyde Tombaugh discovers Pluto in 1930. NASA's New Horizons probe takes off on a nine-year mission to Pluto on 19 January 2006, carrying his ashes with it. On 24 August 2006 the International Astronomical Union issues its papal encyclical: *Resolution 5A: Definition of a Planet in the Solar System.* Six months into New Horizon's mission, the General Assembly of IAU members agree 'that a "planet" is defined as a celestial body that (a) is in orbit around the Sun, (b) has sufficient mass for its self-gravity to overcome rigid body forces so that it assumes a hydrostatic equilibrium (nearly round) shape, and (c) has cleared the neighbourhood around its orbit.' Eight planets, including Saturn, still qualified as such; while Pluto would become a 'Pluto-class object' by virtue of the IAU's new ruling. According to one eyewitness who was present at the General Assembly, the decision had been made quickly and without much protracted debate as it was almost time for lunch. Instead of swinging by the ninth planet in the Solar System, as originally planned, New Horizons would consequently rendezvous in 2015 with a dwarf: one of the larger celestial fragments that lies beyond Neptune in the Kuiper belt. Cosmonaut Third Class Sergei Krikalev experiences a similar change on 26 December 1991 when the Soviet Union is officially dissolved while he is still on a mission aboard the Mir space station. Budgetary wrangles between Russia and Kazakhstan over funding the Baikonur Cosmodrome keep him trapped in orbit for several more weeks. By the time he finally sets foot on Earth again in March 1992, not only does the country that has sent him into space no longer exist, but he also finds that his hometown has changed its name from Leningrad to St Petersburg.

IV

From the middle of the nineteenth century until well into the early decades of the twentieth, it was postulated that another planet must exist somewhere near the centre of our Solar System. This world, whose only tangible presence in the current cosmos was the name 'Vulcan' linked to a set of calculations, appeared to be influencing the behaviour of Mercury, shifting its position in the heavens faster than expected. Apart from the claims of a few astronomers to have seen it, the planet Vulcan never escaped the pages of history or took its place among the star charts. Space can never entirely empty itself. Phenomena that are detectable only through their effects maintain a phantom presence. Percival Lowell of Flagstaff Arizona, who spent the better part of his career arguing for the presence of irrigation canals on Mars, occupied his later years with the search for a mysterious 'trans-Neptunian Planet X', thought to exert a gravitational influence on Uranus.

The light pollution emitted by cities hides the night sky. Stars disappear into the dull halo sprawled out above cities and motorways. Using antennae at the Bell Telephone Laboratories in New Jersey and an obsolete suborbital satellite transmission system, the astronomer Robert Wilson and physicist Arno Penzias detect a mysterious noise evenly distributed across the entire sky that maintains a constant presence throughout the day and night, meaning that it is unlikely to be the electromagnetic noise generated by nearby Manhattan. What Wilson and Penzias have been tracking turns out to be a low-level remnant of the Big Bang: a cosmic echo left over from the massive explosion with which the universe began. This soft hiss is all that remains of the oldest light in existence, emitted when the cosmos was only 380,000 years old. As the universe cooled and continued to expand, this light shifted down to the longer wavelengths of the spectrum. It is picked up today as cosmic background radiation in the region of 2.7 Kelvin, or -270 Celsius. At this temperature subatomic particles start to behave in unexpected and paradoxical ways: at the quantum level, cosmology begins to look more and more like the continual processing and rearrangement of information, its theoretical immortality or its final annihilation.

When ancient civilisations looked up at the stars what they primarily saw was time in operation: they carefully recorded the rhythms and frequencies of celestial occurrence. Astronomy became an art by which to study the mind of God. The Renaissance marked a shift from a time-based astronomy to a predominantly space-based one; the height and distance of the stars revealed space in operation first as geometric form and then as physical force. A similar shift occurred from the use of materials that communicate over time rather than space (clay tablets) to those that favour space over time (paper). The emergence of data-based astronomy marks the convergence of astrophysics with particle physics; the mind of God is transformed into the eternal transfer of information far beyond the narrow bandwidth of our senses. The universe turns out to be one gigantic computer.

V

Poetry is information made aware of itself: the universe cannot exist without it. Projected onto the blank depthless screen of space, images hang like dust in the dark emptiness. Every speck and detail is caught here.

Dante and Virgil stand together, looking down into the Eighth Gullet of the Eighth Circle of Hell: a high-walled cosmos made up of countless tiny flames flickering alone in the dark depths of space. Virgil tells Dante that each flame contains the soul of a sinner. One shakes itself as if struggling against the wind and speaks: '*My brothers, who have confronted a thousand dangers and reached the borders of the west, do not hesitate to follow the sun and explore the uninhabited world.*' It is the voice of Ulysses relating the circumstances of his death. He tells of the time when, old and slow and sailing at the edge of the navigable seas, he urged his crew to turn their oars into wings and, in one mad flight, sail ever further onwards. Always heading south, they set out towards the dawn in search of virtue and knowledge. At night they greeted the stars of the southern pole, while its northern counterpart sank so far below the horizon that it could barely raise itself from the ocean's floor. Having journeyed long and deep over many months, Ulysses and his men rejoiced at the sight of a mountain in the distance, 'the highest I had ever seen' – but their joy turned to grief when a tempest rose from the new land, and the sea closed over them. And Ulysses found himself in the gullet of Hell reserved for those who deceive others by giving false counsel: damned by the words he had spoken to his brothers at the start of their last voyage together.

Which is one way of telling it – based on all that Dante knew at the time from his Latin sources. The cunning and impiety shown by Ulysses will always burn their way through his restless urge to explore. A robotic probe sent out to study solar radiation by the European Space Agency in association with NASA and the Jet Propulsion Laboratory was named Ulysses after the forked flame encountered in Dante's *Inferno*. Between 1994 and 2008 the probe made three complete orbits of the Sun's poles, mapping the storms and winds of the heliosphere. Growing progressively colder over time, its fuel supply coming close to freezing, Ulysses was finally instructed to switch off its transmitter at 20:15 Coordinated Universal Time on 30 June 2009. It is still out there somewhere – a dot in the frozen depths of Hell.

Exploration is the steady colonisation of disaster. Each journey into space ends in the wreckage of entire worlds – how could we ever recognise ourselves in Hell? Astronauts and cosmonauts kiss their amulets and carefully anchor their good luck charms in zero gravity.

Earthly superstitions and rituals serve to generate a tiny cosmos of goodwill around them. The flame that leads us onwards will also mislead. There is no conclusion – only more space.

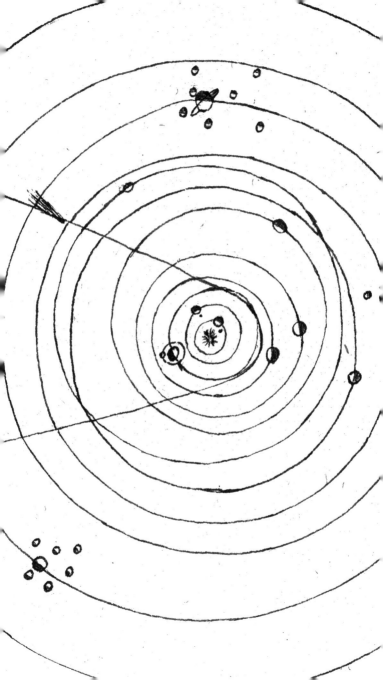

ACKNOWLEDGEMENTS

This book owes its origins to the year I spent with the artist Aleksandra Mir doing research for her *Space Tapestry* project, for which I also received generous funding from the UK Space Agency. During this period I had the privilege of talking with some individuals whose relationship with space has often been more intimate and informed than my own – most notably Marek Kukula of the Royal Observatory at Greenwich, Sanjeev Gupta of Imperial College London, Thais Russomano of the Microgravity Centre in Brazil and Stephen Johnston of the Museum of the History of Science in Oxford. My sincere thanks are due to all of the above for their contributions to the creation of this book. I am also grateful to Rachel Hollings, Richard Barnett, Ross MacFarlane of Wellcome Institute, and Mark Pilkington of Strange Attractor Press for their help in completing the final draft of *The Space Oracle*. Without their encouragement and invaluable insights, I would still be staring into space. The book design and typography of *The Space Oracle* are the work of Tihana Šare, to whom I am greatly obliged. It is, however, much more difficult to acknowledge my debt to the source of the beautiful illustrations that are reproduced throughout this book. They have been taken from a private album of star charts and astronomical illustrations created by an unknown hand sometime over the middle years of the nineteenth century and now residing in the Wellcome archives. The only clue to their identity is offered by a legend written in neat pencil on the fly page: 'Church Lane, Stoke Newington, 1851.' There is a signature in the same delicate line, but it is, alas, impossible to decipher. I hope that this greeting across the centuries from one stargazer to another will serve as some small token of my respect and love for their work.

KH

INDEX

Index

Index

BIOGRAPHY

KEN HOLLINGS is a writer, broadcaster, cultural theorist and lecturer based in London. He is the author of the books *Destroy All Monsters*, *Welcome to Mars* and *The Bright Labyrinth*. His work has been published in numerous journals and anthologies throughout the world, and he has written and presented programmes for BBC Radio 3, Radio 4, NPS in the Netherlands, ABC Australia and Resonance 104.4 FM. Hollings teaches regularly at the Royal College of Art and Central St Martins and has presented his work at the Royal Institution, the British Library, Tate Britain, the Berlin Akademie der Künste, the International Space University in Strasbourg and the Venice Biennale. More information at kenhollings.blogspot.com or follow @ hollingsville on Twitter and Instagram.

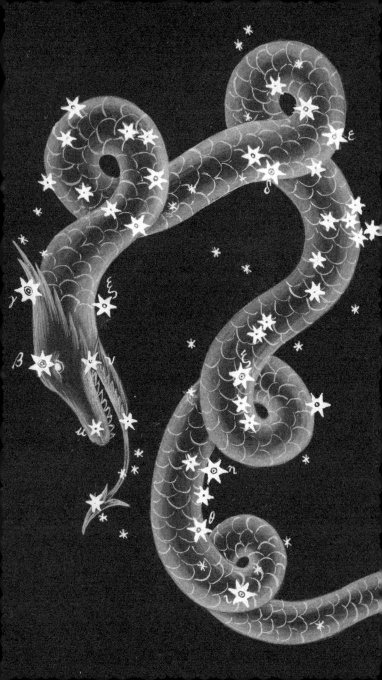

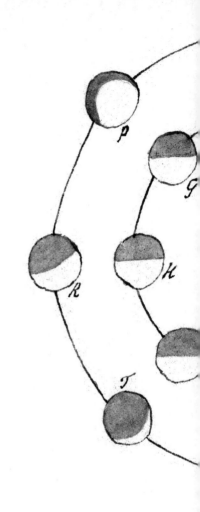

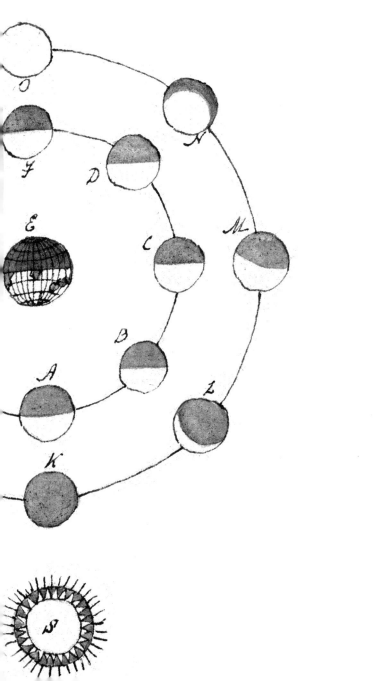

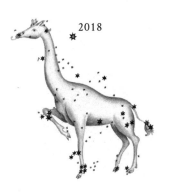

2018

STRANGE ATTRACTOR PRESS